SUN-DIALS AND ROSES OF YESTERDAY

BOOKS BY ALICE MORSE EARLE

CHINA COLLECTING IN AMERICA

CURIOUS PUNISHMENTS OF BYGONE DAYS

CUSTOMS AND FASHIONS IN OLD NEW ENGLAND

SUN-DIALS AND ROSES OF YESTERDAY

TWO CENTURIES OF COSTUME IN AMERICA

Sun-dials and Roses.

SVN DIALS AND ROSES of YESTERDAY

GARDEN DELIGHTS WHICH
ARE HERE DISPLAYED
IN VERY TRVTH
AND ARE MOREOVER
REGARDED AS EMBLEMS

ALICE
MORSE
EARLE

CHARLES E. TUTTLE CO.
Rutland, Vermont

Representatives

Continental Europe: BOXERBOOKS, INC., *Zurich*

British Isles: PRENTICE-HALL INTERNATIONAL, INC., *London*

Australasia: PAUL FLESCH & CO., PTY. LTD., *Melbourne*

Canada: M. G. HURTIG LTD., *Edmonton*

Published by the Charles E. Tuttle Company, Inc.
of Rutland, Vermont & Tokyo, Japan
with editorial offices at
Suido 1-chome, 2-6, Bunkyo-ku, Tokyo, Japan

Library of Congress Catalog Card No. 74-142763

International Standard Book No. 0-8048-0968-2

First edition published 1902 by the Macmillan Co., New York

First Tuttle edition published 1971

PRINTED IN JAPAN

TO MY DAVGHTER

MARY EARLE MOORE

TO COMMEMORATE

HER FIRST SVMMER WITH HER
OWN GARDEN AND SVNDIAL
MAY THE MOTTO
OF HER DIAL
BE THAT OF
HER LIFE

I MARK ONLY SVNNY HOVRS

Table of Contents

List of Illustrations

Publisher's Foreword

The sundials of yesterday were things of deep sentiment. All feel the beauty and wonder of the thought that time—that most fleeting, most wonderful of conditions—was marked so fittingly in its passing by a shadow almost equally intangible, and that the noblest evidences of creation—the stars in their heavens—would be to us invisible and unknown save for their revelation through the shadow of the earth. "Life is a passing shadow," says the Talmud. The sundial points up how irrevocable, how unceasing that passing is.

The same, in a figurative sense, can be said of the rose. But the beauty it brings! Roses were known and loved through the centuries, as they are today. Any book about roses is sure to strike a happy chord in millions of hearts and memories.

It is a pleasure indeed, as well as a privilege, to bring back into print this charming work by Alice Morse Earle and to offer readers the opportunity to enjoy its delights. Certainly, in the present age of haste and waste, it brings true refreshment to the spirit—and a sense of timeless beauty too often difficult to achieve in the world of today.

Foreword

After the publication in 1901 of my book entitled Old Time Gardens, *which contained a chapter upon* Sun-dials, *I received frequent letters (many of them from strangers), asking about sun-dials, their history, construction, manufacture, and cost; about sun-dials in the past, about sun-dials as existing at present, and above all expressing a hope both for sun-dials and a sun-dial book in the future. I found that many of my friends were placing sun-dials upon pedestals in their gardens, or upon the walls of their houses, or wished to erect them as memorials, and were eager to learn of all dials. A general interest in them seemed to have risen in America, as it has ever existed in Scotland and England. As I had for many years collected sun-dials in a desultory manner and information and material in a most assiduous manner, I decided to write this book. Among the sun-dial material were books on dialling, old and new; drawings and photographs of dials; and, not less important, a large correspondence with dial owners, those who possessed single sun-dials and those who owned collections of dials. The union of the subject of Roses with that of sun-dials has not been through any relation of one to the other, but simply a placing together of two "garden delights," — to use Bacon's term, — and with somewhat of the thought that as a dial standing alone in a garden was a bit bare without flowers, so it was likewise in a book. That both are things of senti-*

ment and charm, with something of that magic which in human beings we term fascination, has helped to make their association and companionship in this book a fitting and happy one. I have been aided in the illustration of this book by the thoughtfulness of friends and the generosity of strangers, who have gathered promptly and faithfully photographs of the substantial and permanent beauty of sun-dials as they have the more transient charm of the Rose. To Lewis Evans, Esq., of Russell Farm, Watford, England, I am indebted for lavish illustration and exact information to make my chapter on Portable Dials *a most important record. A. L. Y. Morley, Esq., of Great Brington, England, has brought the historical and beautiful sun-dials of Northamptonshire to me in great variety and number. To Messrs. F. Barker & Son, London, makers of mathematical instruments, and Messrs. William A. Mansell & Co., London, photographers, I am indebted for prompt, intelligent, and faithful assistance, which no business arrangements are ample to reward. To Horace J. Smith, Esq., and Thomson Willing, Esq., both of Germantown, Pennsylvania, I owe the full list of illustrations of sun-dials from Philadelphia and Germantown and vicinity. H. R. Mitchell, Esq., of Haddonfield, New Jersey, and T. S. Robertson, Esq., of Dundee, Scotland, kindly furnished to me many of the drawings and diagrams in these pages. Many have given me single photographs of their sun-dials, or glimpses of their Rose gardens, too many even to name, though I am deeply grateful to each and all. For the chapters upon both the* Sun-dial *and the* Rose as an Emblem, *for those upon the* Rosicrucians *and the* Sun-dial of Ahaz, *I read and studied books, pamphlets, and manuscripts by scores; and, as ever, the treasures of the American Antiquarian Society were of greatest value. In the year 1492—date memorable to*

Americans — there was printed at the monastery at Zzenna a noble volume, The New Psalter of the Virgin Mary. *The border of the first page of the second part of this* Psalter *is a wood engraving of a splendid scrollwork of Rose branches, buds, and blossoms, of bold and almost architectural device. It was cut on oblong blocks, so it could be used in various shaped places. This fine Rose scroll has been adapted, with but slight alteration, as a border for the title-page and dedication-page of this book. The design suggests to our thought the wonderful Rose border of the Kelmscott* Chaucer; *but the Rose sculpture, as it was termed, of the old monkish wood-cutter has more freedom, and in some indefinable way more character than the much bedoubled Roses of William Morris's design. All the fine decorative capital initials which begin the chapters of this book of* Sun-dials and Roses *have been taken from ancient volumes, many of them being appropriately old herbals and books on husbandry. In the early printed books the capitals were designed by artists, but unfortunately their names were seldom preserved. Their work was often grotesque, and even preposterous, but nevertheless (or perhaps therefore) interesting, and above all the initials were always decorative. The works of Erasmus appeared in special luxury of typography, for Erasmus was an intimate friend of Frobenius, the celebrated Basle printer. In a splendid book published in Florence in recognition of Erasmus (dated 1527) is the superb series of initial letters commonly known as* The Playing Boys. *The A of the series is shown on page 426, and is said to be the work of Albert Dürer. A fine example of an heraldic capital is the old black-letter H shown on page 233, from a book printed in Paris in 1514, at the printing-office of one Ascensius. Letters in white upon a darker dotted ground were much used in France, and*

I think the gracefully drawn initial Q in this style of decoration shown on page 87 has a distinctly Gallic touch. For a time the pigment-box of the monk or other limner, painter, stainer, or trickster, often added color or gold to the outlines of the wood-cutter, with glowing results. Toward the close of the seventeenth century the fashion of ornamented capital initials abated. In the following century another taste in capitalization came in; one is shown in the letter on page 318. The chapter on the Rosicrucians *has an initial bearing the mystic symbol of the society. I have taken from these antique books this lovely Rose border and fine capitals to deck this book of* Sun-dials and Roses, *but I cannot see my pages rejoice in the beautiful line edgings, the powderings of gold, the margin-miniatures with which Persian poets adorned their books of Roses, nor can I have the silken paper of flower and fruit tints that they use—pale lemon-yellow, light orange, fine nut-brown, clear iris-blue, and violet, orchid, heliotrope, and lilac, and every tint of Rose, since the glare of white paper offends their eye; nor can I send out my volumes scented with Attar of Roses and Sandal-wood, as were many Persian poems. The copies of the poems of Jami in the Oxford Library, even after four centuries, are fragrant with the original Rose perfumes. Both Sandal-wood and Attar of Roses are far too costly to be used by modern publishers. Once Sandal-wood was free in some amount in Persia to all save beggars; even an historical author could have Sandal-wood gates to his waiting-room. But he found that the perfume so filled his brain that it diverted him from serious thoughts and composition, and made him liable to "drop into poetry"; and he sternly had the fragrant portals removed to his harem, where bemused brains did not matter. In our own day we find a case grotesquely analogous. Walter Savage Lan-*

dor, when seated at his desk, fully prepared and eager to write, would be so diverted from his intent by the scent of the wood of his freshly cut lead-pencil that he would sit for hours, motionless, sniffing the Piny odor, writing nothing. But Pine or Sandal-wood, or any other fragrance, is little heeded or valued to-day; and Attar of Roses is so blended and degraded that we scarcely know the pure Rose perfume.

ALICE MORSE EARLE.

October, 1902,
Brooklyn, New York.

Sun-dials and Roses of Yesterday

CHAPTER I

THE CHARM AND SENTIMENT OF SUN-DIALS

"A Dial is the Visible Map of Time, till Whose Invention 'twas follie in the Sun to play with a Shadow. It is the Anatomie of the Day and a Scale of Miles for the Jornie of the Sun. It is the silent Voice of Time and without it the Day were dumbe. . . . It is ye Book of ye Sun on which he writes the Storie of the Day. Lastly Heaven itself is but a generall Dial, and a Dial it, in a lesser volume."

— *Heliotropum Sciothericum*, Robert Hegge, 1630.

THERE are in nature some simple expressions of usefulness which have a charm that is impossible to define. This charm seems to consist in the direct, the unadorned, and unencumbered application of shape and form to the reason of their being. They are often primitive objects, sometimes those of ancient races, where each line has been shaped out unconsciously through centuries of use, not with any thought of beauty, but to serve distinctly and simply the purpose of existence. Such objects are the snow-shoe and the canoe of our North American Indians; absolutely perfect in outline, devoid of all superfluities, impossible of improvement, they possess in full not only beauty, but the charm to which I refer. An ancient Greek lamp is another example; this classic form of lamp was used not alone in ancient Greece, but in scores of other lands, by mediæval races, and even in humble homes by our own contemporaries. The iron "betty-lamps" of our New England grandmothers, found still in remote New England homes, — lamps with hanging chain, small oval body and protruding lip to hold a primitive wick, differ not in single detail or outline from the lamps of ancient Rome and Egypt. Household lamps retained this antique useful shape as long as the same domestic mediums of illumination were used, — namely, household grease and oil. With the introduction of more lavish means of illumination came varied forms of presentation of artificial light, and the old simplicity of outline of the hanging lamp vanished.

The sun-dial is another striking example of the charm of simplicity in form and directness in utility; its lines and markings are the absolute mathematical expression of the information it gives; it is set on a decorative pedestal or fixed with ornamentation on a wall simply for convenience of our sight. You may elaborate the lines of the dial-face, and decorate the mounting of the dial, but that does not add to its subtility of charm. You feel that curious inter-

Sun-dial at Glamis Castle, Scotland.
Seat of Earl of Strathmore. *Ante* 1620.

est and "drawing unto" in the simplest sun-dial of dull metal fixed on the kitchen window-sill of the humble farm-house, just as you feel it in the wonderful dial of Glamis in old Scotland.

This exquisite monumental dial, deemed by many the masterpiece of all dials, may well open the series of illustrations of sun-dials in this book. It stands on the grounds of Glamis Castle, home of tragedy, legend, and romance; even its picture speaks to us of Macbeth, the shadowy Thane of Glamis, and of the charm and magic of Shakespeare's play.

This picture of the dial is better than any description; but it may be noted that the twenty-four facets of the head have each three and some four dials, giving over eighty dials in all. The rampant lions each hold a fine vertical dial, one of which is elliptic in shape, nineteen inches long; two are square, thirteen and one-half inches in diameter; and the fourth is rectangular and is fifteen and one-half inches long. The lions are separated by four beautiful twisted pillars carved in the spiral hollows. The height of the dial is thus divided:—

Height from ground to place on which the lions stand	3 ft. 7 in.
Height of lions	5 ft. 2 in.
Cornice	1 ft.
From top of cornice to upper part of faceted head .	3 ft. 3½ in.
Facet head	3 ft. 5½ in.
Scrolls and coronet	4 ft. 9 in.
Total . . .	21 ft. 3 in.

The width of the octagonal lower step at its base is ten feet and ten inches; it forms thus, as may be plainly seen, a grand monumental dial, fit for the majestic castle beside which it is reared.

This castle is the residence of the Earl and Countess of Strathmore; and this fine photograph was taken by Lady Maude Bowes-Lyon for this book. The dial is certainly three centuries old, as it appears in a print of the castle previous to the year 1600, and was named in Earl Patrick's *Book of Record* of a date previous to 1695.

Sun-dial at Balcarres Castle, Fifeshire.

The sun-dial has for us another charm — one that is common to all deeds and instruments that note the passing of time. In the days of childhood we gathered eagerly the downy seed-balls of the Dandelion, and as we held them aloft we blew upon them with strong young lungs, and called out: "What's the hour-o'-the-day?" Thus do all children of all lands wherever the Dandelion blows and turns "quite old and gray." Vague was the answer of the bared stem of the Dandelion to us; and I doubt if we expected to learn from it the time. It might have answered in nearly the words of the old sun-dial motto: *Hora non numero nisi juventas.* I COUNT ONLY YOUTHFUL HOURS. We ask the hour with equal intentness of the long-legged garden-spider and of the grasshopper:—

"Grandfather, grandfather gray,
Tell us the time-o'-the-day."

We had thus early in life the universal instinct of humanity, a longing to count the hours and minutes of passing time; and we never wearied of the trial. How full of significance also is the hour-glass, how classic its shape; what a charm has it for the child—just as it had in the childhood of life for ancient peoples.

With what exquisite perfection of simplicity has Tennyson, in his *In Memoriam*, characterized the succession of marking the passing of Time by hour-glass, sun-dial, and clock!—

"For every grain of sand that runs
And every space of shade that steals
And every kiss of toothed wheels
And all the courses of the sun."

The sun-dial is a creature of equal sentiment and sense. Its good sense is proven by its being so perfectly satisfying, so absolute. You may deem its sphere a restricted one, its message a short one; but it fulfils its duty, and tells its story to perfection—

it is satisfying. And this is no small thing when we recall how few objects there are in this world, whether formed by nature or shaped by art, that are satisfying. Try to name them! the perfect, the wholly satisfying things you know; there a few books — and alas! how few they are; and some pictures — I can count them far too quickly. Roses and Fritillaries are as absolutely satisfying as the sun-dial, and happily many trees. A Ural Mountain amethyst, yes, and two cocker-spaniels, friends of mine; once in a lifetime a gown; I suppose architects could name some buildings for this list; and some folk may have had a perfect horse; and I know a few perfect pieces of domestic furniture, of silver, of china. But nearly all sun-dials please us absolutely — certainly all simple and direct ones, and I think it well worth while to exist merely to be satisfying if nothing more.

Sun-dial at Kelburne House, Ayrshire; Seat of the Earl of Glasgow.

But the sun-dial is a thing of deep sentiment.

All feel the beauty and wonder of the thought that Time, that most intangible, most fleeting, most wonderful of conditions, is marked so fittingly in its passing by a shadow almost equally intangible; and that the noblest evidences of creation — the stars in the heavens — would be to us invisible and unknown save for their revelation through the shadow of the earth. Thus are great truths revealed to us, not by great Light but by Darkness — a lesson of Life.

The Quaker poet Bernard Barton felt the sentiment of the sun-dial; it accorded well with his temperament and his faith. Here are his noble verses: —

"With still more joy to thee I turn,
Meet horologe for Bard to love;
Time's sweetest flight from thee I learn,
Whose lore is borrowed from above.

"I love in some sequestered nook
Of antique garden to behold
The page of thy sun-lighted book
Its touching homily unfold.

"On some old terrace wall to greet
Thy form and sight which never cloys;
'Tis more to thought than drink or meat,
To feeling than Art's costliest toys.

"These seem to track the path of time
By vulgar means which man has given.
Thou — simple, silent, and sublime —
But shows thy shadowy sign from Heaven."

"Simple, silent, and sublime" — in its silence the sun-dial is strong.

There is such severity, such dignity in the noiseless marking of the flight of Time; no irritating ticking, no striking of the hours, no sounding of bells. Silently as surely, the hours pass away, and the day with even measure balances its periods till the setting sun leaves a darkness equal to the silence.

"The sly shadow steals away upon the dial and the quickest eye can discover no more, but that it is gone," wrote Glanville. There is an element of mystery in this imperceptible flight, and all mystery is alluring; you may note the swaying pendulum of the clock, or you may hear the ticking of the watch, you may see the tiny stream of sand of the hourglass, but you can see no movement of the shadow; "nice," said Lamb, "as an evanescent cloud or the first arrests of sleep."

How vast, how wonderful is the thought of Life, of the passing of Time! How crude, how paltry our definitions! How petty our explanations! Only by symbolism can these things be expressed!

In the Talmud are these fine lines:—

> "Life is a passing shadow, says the Scripture. Is it the shadow of a tower? of a tree? a shadow that prevails for a while? No, it is the shadow of a bird in his flight—away flies the bird, and there is neither bird nor shadow."

We cannot hold this shadow, if we would, but its passing is shown to us on the sun-dial. And on the dial-face alone does this passing seem irrevocable—unceasing. You may refuse to turn the hourglass and thus deceive yourself that Time flies not. You can cease to fill the water-clock and let the weights of your clock run down until its hands turn

Sun-dial in Walled Garden, Wratisbury, England.

not — but you cannot check the course of the sun's shadow.

Wonderful as is this thought of the present of the dial, its past is more profound.

> "The shadow on the dial's face
> That steals from day to day
> With slow unseen, unceasing pace
> Moments and months and years away,
> This shadow which in every clime
> Since light and motion first began
> Hath held its course sublime."

"Since light and motion first began": when on the Fourth Day of the Creation, God said, —

> "Let there be lights in the firmament of the heaven to divide the day from the night, and let them be for signs, and for seasons, and for days, and for years.
>
> "And let them be for lights in the firmament of the heaven, to give light upon the earth; and it was so."

And the dial *was so* also; the trunks of the trees were gnomons, there was light, there was motion, there were shadows, and therefore there were sun-dials. As Charles Lamb said of a dial, "Adam could scarce have missed it in Paradise."

Certain inanimate objects have a semi-human closeness to us. I do not by this refer to objects with which we have intimate and happy associations, such as a chair in which loved ones have sat, a desk at which we ourselves have long written; but I mean that an inherent quality is possessed by some objects which even at first sight makes them seem almost human. I always feel this quality in mile-stones and in old windmills; of course we all have known deep attachment for certain books, which is natural enough, since they have spoken so plainly to us. Many musicians know this feeling for and

Moot Hall, with Sun-dial, Aldeburgh, England.

about their musical instruments, and workmen often have it and always should have it for their tools.

Many feel this with clocks and watches, and I am deeply sensible of it in a sun-dial. Of course, in the dial, it may be partly because the dial has a voice, its lines and numerals speak if it has no motto; but it should always bear a motto or inscription. This at once gives it a nearness to humanity; it is our kinsman, our fellow-countryman; it speaks our language. The pointing hand of the guide board gives to it a semi-human appearance; the simple words of the mile-stone make us ever interested in it; all inscriptions draw us close to the thing inscribed. I have told often of my love of mottoes, legends, inscriptions, notes inscribed everywhere.

We have an orchard seat, and such a seat among fruit trees has a fresh pleasure for every spring morn and summer day. On the yellow pine surface that forms the back of this seat, a friend has lettered in heavy ink, which we renew in blackness each spring, these lines from Wordsworth, which the poet might have written with this very orchard spread around him:—

> "Beneath these fruit-tree boughs that shed
> Their snow-white blossoms on my head
> With brightest sunshine round me spread
>   Of Spring's unclouded weather;
> In this sequester'd nook how sweet
> To sit upon my orchard seat!
> And flowers and birds once more to greet,
>   My last year's friends together."

Now what is the result of that inscription? It is this: the commonplace orchard seat was made at once a different being; it was given a voice—and that voice was the voice of a friend. It did far more than to speak to us of the friend who transcribed the lines; it brought to us Wordsworth, and his orchard seat, and then the beauties of the Lake Country; it made travellers think of the birds seen there; and it spoke to us of many old friends who had sat with us in the orchard. Thus it is with the motto on a sun-dial; it ever speaks, a different message perhaps to each who reads it, but an inspiring message, one sometimes of great moment.

A motto of wonderful power is the few words from the New Testament, FOR THE NIGHT COMETH. In Greek, Hebrew, Latin, or English this motto is seen. To the thoughtful mind it ever recalls the solemn scene where the warning words were spoken, our Saviour's admonition to prepare for eternity. It spoke with infinite force to Sir Walter Scott when he read it on his dial at Abbotsford, urging him to incessant work. The story of his dial is told in his *Life* by Lockhart, and the curious fact that the Greek words of its inscription were incorrect. The presentment of his dial is shown on page 13; the photograph was not taken from the original dial, but from an exact reproduction of it in the garden at Hillside, Menand's, New York. The original dial was sadly worn and disordered when it was drawn for Mr. Douglas the publisher. He had it repaired and reset, and had this reproduction made. It is exact as to lettering as well as shape, impresses having been taken from the dial.

Another thought comes forcibly in the words, FOR THE NIGHT COMETH, — the absolute cutting off of all power of marking the passing of time through the shadowing of the dial by night. It is an im-

Sun-dial at Hillside, Menand's, New York.

pressive thought, — the death of a day. Rossetti thus expressed it: —

> "Slowly fades the sun from the wall
> Till day lies dead on the sun-dial."

The sentiment and beauty of the sun-dial appealed to and charmed many a poet. I have gathered in my Common-place Book a florilege of scores, almost hundreds of verses, relating to the sun-dial. Some of these sentiments are most tender and touching; and with the spirit of most of them I can sympathize. I give the opening verses of lines written by Hugh Miller to show the notions he wished to express, though they convey not a single word of my thought of a sun-dial:—

"Gray dial-stone, I fain would know
What motive placed thee here
Where darkly opes the frequent grave
And rests the frequent bier;
Ah, bootless creeps the dusky shade
Slow o'er thy figured plain:
When mortal life has passed away
Time counts his hours in vain.

"I think of those that raised thee here,
Of those beneath thee laid,
And ponder if thou wert not raised
In mockery o'er the dead.
Ah, never sure could mortal man,
Whate're his age or clime,
Thus raise in mocking o'er the dead
The stone that measures time."

There still stands at the old home of Hugh Miller an ornate dial-stone (it will be noted that he never says sun-dial) which he cut for amusement in a period of recovery from illness; it is near another dial, an ancient one which he dug out of the earth when he was a boy, and which had originally been set up in the old Castle garden of Cromarty. By the side of this ancient dial Miller first saw the young girl who afterward became his wife. The

Angel with Sun-dial on Cathedral, Genoa, Italy.

dial-verses were written in early youth; an age when most poets love to write upon death and gloomy moral lessons. Perhaps had he written them after he met his sweetheart, they might have been more natural. However, the chief reason why I do not like them is that they are not poetry; they form a perfect example of Dr. Edward Everett Hale's amusing method given in his advice *How to Write*, an exercise of "capping verses."

A true lover of Charles Lamb asserts that he ever finds in Lamb the best thoughts on any subject — whatever it may be; thus, upon sun-dials he would believe that the ideal sentiment was expressed by Lamb in his *Essay on the Old Benchers of the Inner Temple*. It is indeed inexpressibly fine in poetic feeling — far beyond any poem we have; and deserves quotation in full by all who write on dials: —

> "What an antique air had the now almost effaced sun-dials with their moral inscriptions, seeming coevals with that time which they measured, and to take their revelations of its flight immediately from heaven, holding correspondence with the fountain of light! How would the dark line steal imperceptibly on, watched by the eye of childhood eager to detect its movement, never catched, nice as an evanescent cloud, or the first arrests of sleep!
>
> "Ah! yet doth beauty like a dial-hand
> Steal from its figure, and no pace perceived.
>
> "What a dead thing is a clock, with its pondrous embowelments of lead and brass, its pert or solemn dulness of communication, compared with the simple altar-like structure and silent heart-language of the old dial.

"It stood as the garden god of Christian gardens. Why is it almost everywhere vanished? If its business use be suspended by more elaborate inventions, its moral uses, its beauty, might have pleaded for its continuance. It spoke of moderate labours, of pleasures not protracted after sunset, of temperance and good hours. It was the primitive clock, the horologe of the first world. Adam could scarce have missed it in Paradise. It was the measure appropriate for sweet plants and flowers to spring by, for the birds to apportion their silver warblings by, for flocks to pasture and be led to fold by. The shepherd 'carved it out quaintly in the sun,' and turning philosopher by the very occupation, provided it with mottoes more touching than tombstones."

Dante's Amor.
By Dante Gabriel Rossetti.

I have ever been struck with one expression of Lamb in writing of the sun-dial; he called it "a simple altar-like structure." It is partly the classic shape of the sun-dial — its altar-like form — which charms us; and a proof to me of the wisdom of simplicity in outline for every dial-pillar is in the fact that the simpler forms evoke the greater sentiment.

I find that half the folk who speak of sun-dials like to quote Austin Dobson's verses on a sun-dial, and worthy of quotation they are, and full of sentiment: —

"'Tis an old dial dark with many a stain.
In Summer crowned with drifting orchard bloom,
Tricked in the Autumn with the yellow rain,
And white in Winter like a marble tomb.

And round about its gray, time-eaten brow
Lean letters speak — a worn and shattered row: —
'I am a Shade — a Shadowe too, art thou.
I mark the Time. Saye! Gossip! Dost thou soe?'"

The last couplet has been used as a motto on several sun-dials both in England and America. On a dial at Grey Friars Churchyard, Stirling, is a similar motto: —

I AM A SHADOW, SO ART THOU.
I MARK TIME — DOST THOU?

Rossetti felt deeply the significance and charm of the sun-dial. He wrote these beautiful lines: —

"Stands it not by the door?
Love's Hour — ?
Its eyes invisible
Watch till the dark thin-thrown shade
Be born, — yea, till the journeying line be laid
Upon the point that notes the spell."

What mystery the presence of a dial adds to his beautiful painting of *Beata Beatrix*, where a horizontal dial on the widow-sill marks to Beatrix the coming of her wonderful death-trance. On page

16. I have given a reproduction of the angel in Rossetti's beautiful pencil sketch called *Dante's Amor*. This angel holds an ancient Saxon sun-dial. Many of the cathedrals on the Continent have carved angels on brackets or corbels holding sun-dials. A beautiful angel with dial at Chartres is here shown; also facing page 14 a still older carving upon the Genoa cathedral. These figures offer wonderful suggestion for a memorial window-dial, such as is described in the succeeding chapter.

Angel with Sun-dial, Cathedral, Chartres.

As an object of interest and romance in a garden, the sun-dial has a strong hold on our sentiment; we have seen that artists have painted it and poets have written of it. As a mystery to childhood, a trysting-place for faithful lovers, a sad reminder to a deserted sweetheart; a subject for moralizing for the preacher, and of reminiscence to the aged gardener, its place in pictures — either in print or on canvas — is a permanent one. Of all spots for a garden-dial the focus of a formal garden is the most suited; that focus may be the centre, or where several paths converge, or in a recessed end; but wherever it is, the dial should be the point of high interest. From its very nature it is (unless miserably hidden) that point of interest. The poetical suggestion of a sun-dial never could be more fully shown than in the fine picture opposite page 20 of dial and man; for the man is the great English artist, George F. Watts, who has given to us a conception of the passing of Time and of Death which has ennobled Art, and robbed Death of its horror. And it is a beautiful thought that his dial bears the motto of the artist's life — THE UTMOST FOR THE HIGHEST. I can never adequately express my gratitude for the kindly and thoughtful gift of this photograph taken solely with intent to gratify an unknown author across the seas, through the timely sending for the illustration of her book, this counterfeit presentment both of artist and dial. This dial, with its faceted head of antique design, was made for Mr. Watts at the Arts and Crafts Association of his own village — a village industry where modelling in terra cotta is taught and done.

Through its inherent characteristics of picturesqueness, symbolism, and sentiment, the day of the dial in England has been a long one; but in our new world we have not always regarded sentiment in our surroundings, and sun-dials have been in retirement. In our cities they have vanished. I did not for years, until about six years ago, know a sun-dial upon a building in New York or Brooklyn, save one a stone's throw from my own home. By the side of Grace Church, in Brooklyn, there runs down to the open gardens on the beautiful Heights which overhang the wharves of the harbor, a short and quiet street called Grace Court. Dwelling-houses are built facing the north on one side of the Court, while on the other side are no houses; and there is a fine expanse of adjoining gardens in the rear of the row of houses which face on Remsen Street—an unusual expanse for city streets. In those gardens and around the church there lived in our crowded city, from early spring till midwinter, with life as free as in his native wilds, a great white cockatoo, who had escaped from some South American ship as it lay at the wharf under the Heights. Hiding in the trees in the daytime, and perhaps in the church tower, he tapped at friendly windows at night, like a white-winged ghost, confident of the welcome and food which he always found; sometimes he screamed out harshly in angry hunger, and sometimes he spoke, as he tapped, foreign words of greeting or comment taught him by the sailor who had brought him to this port.

Into one of these gardens stretches out an artistic two-storied extension of fine brick and terra-cotta walls; and in the apex of the gable, facing the direct south, is a large bronze sun-dial of triangular shape. It can from its prominent position be plainly seen by passers-by and church attendants; and it has

George F. Watts, R.A., seated in his Garden
by his Sun-dial. Limnerslease, Guilford, England.

been a never ceasing source of pleasure to me for many years to note how closely it corresponds to clock time. I never fail to glance at it in passing. I used to hope to see the cockatoo wisely bending over it; and that he would croak out to me over the gardens, "What's the time-o'-the-day?"

The owner of this sun-dial when so few cared for sun-dials, and many had never seen one, was Samuel Bowne Duryea, Esq.; and he put his fancy for sun-dials to practical use, laying one around the flag-staff at Robin's Island, tracing the analemma in colored stones, where it was an object of interest to all who saw it.

It is strange that the sun-dial should have been so generally neglected; our patriotism should have made us cherish it as an emblem closely connected with the early material prosperity of the United States. I have told at some length in my book entitled *Old Time Gardens*, of the interesting presence of the sun-dial in our national history; but I must refer to it again here. In the first coinage of the United States a sun-dial made frequent appearance. A design of a sun-dial was on the dollar which was cast in silver, then in bronze, then in pewter; it appeared on the copper cent and was printed on a paper note of the value of one-third of a dollar. This sun-dial bore two inscriptions, one *Fugio*, the other, MIND YOUR BUSINESS. The word *Fugio* gave a name to this currency, and the pieces were known as the "Fugio dollar," the "Fugio cent," and the "Fugio note." The cent was also called the "Franklin cent," and is so known by collectors to-day. This was through

Franklin's connection with the coinage. The "Fugio note" is here shown, also a later use of a similar sun-dial design on a local note issued by the city of New York.

It will be recalled that Franklin had known much of the postal service of Great Britain before he

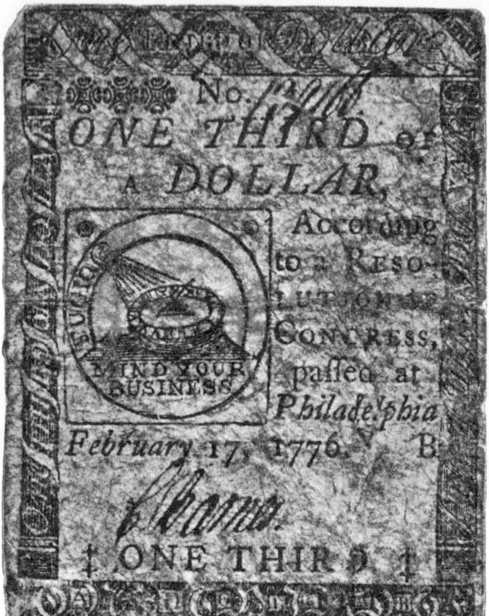

Fugio Note.

became postmaster-general for the American colonies under the crown. And he had lived long in London, where on the general post-office was a sun-dial with the motto, BE ABOUT YOUR BUSINESS. I have never doubted that it was entirely Franklin's taste which supplied to our new nation the sun-dial design and the motto, MIND YOUR BUSINESS. In this form, and the one on the London post-office, and in the form, BEGONE ABOUT YOUR BUSINESS, it was found on several English sun-dials. The one in the Inner Temple owed its motto to a surly reply given to a dial-maker who asked at the Temple library, as he had been instructed, for "the motto for the new sun-

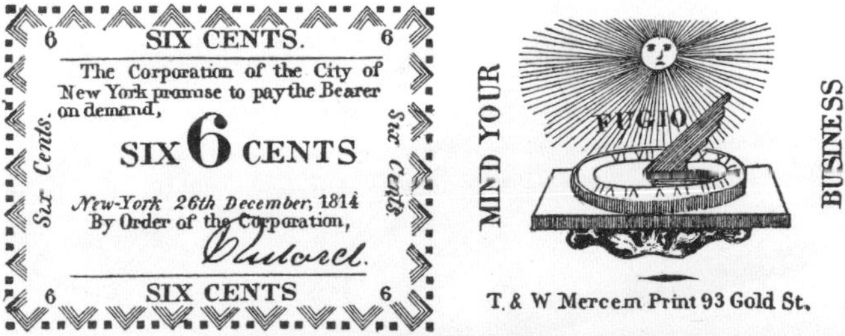

Six Cent Note of City of New York,

dial, sir." "Begone about your business!" was the testy answer of the only inmate of the library. And a very good motto it seemed to the dial-maker, and the Benchers also, after it was put up.

In the eighteenth century dials were an article of common manufacture in America, though I think never in large numbers. Seldom do we find them named in old tradesmen's lists. I have seen fifty-eight different articles enumerated in one pewterer's list, but sun-dials were not among them. Perhaps the fact that each dial was limited in its sphere — could not be used save in its own latitude — hindered their production. In England and on the Continent people lived in close-lying towns; in England the variation of latitude could not be vast; but in the new world all was different. Distances were great. And those distances were chiefly in latitude, — up and down the coast. Therefore, portable dials would be sought rather than fixed ones. There still exist in America, however, old soapstone moulds used for the casting of pewter sun-dials.

Ellicott Sun-dial.

The steatite mould of George Ellicott, of Bucks County, Pennsylvania, is in good condition. He was a well-known maker of sun-dials and high case-clocks, — a son, I believe, of the engineer Ellicott who did so much of the laying out of the Federal city of Washington, and the District of Columbia. Here is a pewter dial with base and gnomon in one piece made recently in this cast. It is about five inches square; is marked "1779 G. E." The hours are in Roman numerals and "Lat 40" is on the side of the gnomon. I own a much-worn pewter dial with circular base, bearing the same initials and date. It was given me by a friend who purchased it in Bucks County and paid for the tradition that it was made by Ellicott, as it undoubtedly was.

I find that to many the sun-dial is an emblem and voice of some great sentiment or hope in life, or a distinct reminder of some scene or incident, often of childhood. Let me tell the story of Harriet Martineau's sun-dial as an example; it can best be given in words from her *Autobiography*. She went with her sisters and brothers when she was seven years old to visit her grandfather. On the way thither the five children were amused by being told to guess what they would find standing in the middle of the garden. On her arrival, rudely ignoring the happy welcome of the tearful old people, she insisted on seeing "the thing in the garden." She writes: —

"I could make nothing of it when I saw it. It was a large heavy stone sun-dial. It is worth this much mention for it was of immeasurable value to me. I could see its face only by raising myself on tiptoe on its step; and there, with my eyes level on the plate, did I watch and ponder, day after day, painfully forming my first clear conceptions of Time amidst a confusion of notions of day and night, and of the seasons, and of the weather. I loved that dial with a sort of superstition, and when nearly forty years after, I built a house for myself at Ambleside, my strong wish was to have this very dial for the platform below the terrace. But it was not to be had."

Another dial, however, she did have, and the story of its setting up runs thus, in her words: —

"A friend in London who knew my desire for a sun-dial and heard that I could not obtain the old one which had told me so important a story in my childhood, presented me with one to stand on the grass under my terrace wall and above the quarry which was already beginning to fill with shrubs and wild flowers. The design of the dial is beautiful — being a copy of an ancient font; and in grey granite, to accord with the grey-stone house above it. The motto was an important affair. A neighbour had one so perfect in its way as to eclipse a whole class, — the class of Bible-sayings about the shortness of life and the flight of time, — 'The Night Cometh.' In asking my friends for suggestions, I told them of this, and they agreed that we could not approach this motto in the same direction. Some good Latin ones, to which I inclined, were put aside because I was besought, for what I considered good reasons, to have nothing but English. It has always been my way to ask advice very rarely, and then to follow it. But on this occasion I preferred a motto of my own to all that were offered in English; and Wordsworth gave it his emphatic approbation. 'Come! Light! Visit me!' stands emblazoned on my dial; and it has ever been, I believe, as frequent and impressive a monitor to me as ever was any dial which bore warning of the fugacious nature of life and time."

I think no one can read these fine and forceful extracts without feeling a deep interest in this dial, and I am glad to present here the artistic photograph of it sent me by Miss Martineau's niece.

The sun-dial in the garden of Sir William Humphrey, Great Brington, Northamptonshire, has the same speaking motto, COME LIGHT! VISIT ME. The great beauty of this dial-pedestal, and its lovely setting of tuberous Begonias is shown on page 29.

Sun-dial of Harriet Martineau, The Knoll, Ambleside, Westmoreland, England. 1846.

To me the sun-dial ever recalls two incidental scenes in my life. The first, through some curious psychological twist, is one in which a sun-dial took no part whatever; it was the only time in my life when I felt alone in the world.

To few people and but seldom is it given to feel utterly alone with nothing but the sun and the earth. Richard Jeffries, in that perfect prose poem *The Story of my Heart*, tells of the pantheism of the hills; of his sense of loneness on a hilltop, that the earth held him and pressed him and spoke to him, and he felt an emotion that was as if his whole life were poured out in a prayer. It was in midsummer that a similar sense came to me as to that strange creature, Emily Dickinson: —

> "There came a day at summer's full
> Entirely for me ;
> I thought that such were for the saints,
> Where revelations be."

I had driven with my father to a remote farm, and we had gone into a half-evergreen pasture to gather from the abundance of exquisite Azaleas, when my father recalled that he had left an overgarment at the empty farm-house adown the hill, and he drove back to secure it, leaving me alone flower-gathering in the rocky hill-pasture.

There was not a house in sight, for an edging of fine old pine woods surrounded the pasture, and the tall tree-spires cut it off from the rest of the world and left it high on the hilltop, and the road thither was scarce more than an overgrown lane and soon vanished into the trees. I had as I sat there waiting, a distinct impression, as did Jeffries, that I was alone in the world. My father would never return, I should never see mortal face again; and I did not care to. I was so filled with the beauty of the scene, the perfume, the song of birds, above all the great heat and glow of that radiant sun of June that I was possessed with a sort of obsession; an absolutely pagan sense of sun-worship and of the isolated completeness of that beautiful moment — and I felt no desire for life beyond, either in this world or the next; though, as the old poet Vaughan said, —

> "I felt through all this fleshly dress
> Bright shoots of everlastingness."

All my thoughts and senses seemed emancipated. I was conscious to the keenest degree of what Shakespeare termed "my glassy essence." I cannot, of course, feel thus whenever I stand by a sun-dial; but the dial always recalls this scene to my mind. It speaks with no uncertain voice of that afternoon when I was alone in the New England hill-pasture and in the whole world.

The second scene is not so remote in my life; it was nearly fifteen years ago that I was shown a friend's sun-dial; one of the few garden-dials then to be found in America. I saw it on one of those strangely warm and beautiful days which we have sometimes during the first weeks in April in New England, — an April which is often bleak as the first of March, and not wholly absolved from dread of snow flurries. These beautiful days of April are like none other; for the sun is so burning at mid-day and there is such a pervasive feeling of tentative greenness, though nothing is really green. I have shown this atmosphere to a wonderful degree for black and white in an illustration on

Sun-dial of Sir William Humphrey, Bart., Great Brington, Northamptonshire.

page 155 of my *Old Time Gardens* in a Lilac picture entitled *Opyn-tide, the Thought of Spring*—"Whenne that flowres think on blowen." On such a day we suddenly find that there are Ladies' Delights in bloom as well as Snowdrops, and the quick ear catches the buzz of bees, — ever welcome and happy sound after months of snow and silence in the garden. There is an old Chinese saying, —

> "Ere Man is aware
> That the Spring is here
> The Flowers have found it out."

And we can add, "The bees have found that the flowers are out." On this day in Opyn-tide we followed the loved sound of these humming bees around a Lilac corner, and there they were, surrounding the sun-dial, bumping against it in their heavy, benumbed first flight. And there — open so wide in the hot sunshine that their glowing petals seemed fairly reflexed to the base of the stems, not only open but bent back to drink in the sunshine — were scores of beautiful purple and golden and snowy Crocus blossoms, planted in affection that the sun-dial might have the first flowering of spring.

There, by the sun-dial and the shining Crocus-cups, came to me a line of rare Ben Jonson's, —

> "The World may find the Spring by following her,"

a line which might have been written for my mother. With such inner light did she know where flowers grew — whether in garden, grove, or meadow, — so constantly was her path filled with flowers, that they seemed to throng lovingly around her rather than that she went to search for them.

> "Here was she wont to go! and here! and here!
> Just where these daisies, pinks, and violets grow,
> The world may find the Spring by following her.
> And where she went, the flowers took thickest root."

She ever gathered in gentle triumph the firstlings of spring,—the earliest Snowdrops, the little stunted Ladies' Delights, the half-frozen spires of Scilla. And she ever spied, ere we deemed them out of the frozen ground, the first glint of yellow Crocus.

So as I stood by this dial I had a picture in my heart; the one which ever comes to me now as I stand by a garden-dial. I could see my mother's eager, bright-eyed, smiling face, as she leaned over the Crocus bed and listened to the murmurous hum of the bees as they buzzed, half-chilled, from flower to flower. "How do they know that winter is gone," she said, "when we scarce know it ourselves? Where have they been throughout the snow? From whence do they come? How do they know—who told them—that here in my garden these purple and yellow cups are opened for them?"

* * * * * * *

> "Ah, far away in some serener air
> The eyes that loved them see a heavenly dawn,"

and I sigh as I turn from the sun-dial, but I read its motto: *Lux et Umbra Vicissim, sed semper Amor*—LIGHT AND SHADOW BY TURNS, BUT ALWAYS LOVE.

CHAPTER II

NOON-MARKS, SPOT-DIALS, WINDOW-DIALS

"The learned line showeth the city's hour."
—Motto on Sun-dial in Milan.

"Little sun upon the ceiling
Ever moving, ever stealing
Moments, minutes, hours away.
May no shade forbid thy shining
While the heavenly sun declining
Calls us to improve the day."
—Motto on Ceiling-dial at Theobald's.

"Whilst Phœbus on me shines
Then view my shades and lines."
—Motto on Manz Dial.

I WONDER whether you, my gentle reader, have ever read a book entitled *Margaret*, which was published just half a century ago. Its author was Reverend Sylvester Judd, a New England minister of severest Puritan rearing and environment. He says in the curious "author's note" which forms a preface, that he spent in writing it over ten years; meaning by that the hard-won hours of leisure of a decade

The Oldest English Sun-dial, in Churchyard, Bewcastle, England. 690 A.D.

of the life of a New England "painful preacher." The first part of the book offers, without doubt, the most valuable picture which exists of domestic life in a small New England community in the years that "take our country as it emerges from the Revolution," and for half a century thereafter; not a grandly rounded picture as a whole, but a photographic presentation of details. On the singular charm of the book I do not dwell, as all are not sensible of it. The author says in his fanciful "phantasmagorical" first chapter or introduction:—

> "The child is MARGARET, of whom we have many things to say, and whom we hope to reveal more perfectly to you. So far as this book is concerned, she is for you all as much as if she were your own child; and if you cared anything about her when you did not know her, we desire that your regards may not subside when you do know her, even if she be not your own child; and we dedicate this memoir of her to ALL who are interested in her and care to read about her."

By the engaging simplicity of this introduction a spell is thrown on many readers. The author said in a later edition that he had been called "unequal, grotesque, mermaiden, abrupt"—and he was called so with truth, though I scarcely am sure what his adjective "mermaiden" means; the book is certainly whimsical and capricious, for the last part portrays as unnatural a picture of life as the first two parts are true. It is all quaint, however, in the truest sense of that (of late) overworked term. The book always seems to me to have been composed under a certain inspiration, an obsession of some spirit of the past. Of course the fact that it was ten years in the writing thereof would not carry out my theory; we always think of true inspiration in the form of a *tour de force*. As a valuable depository of ancient words, phrases, usages, and things, of terms and expressions of colonial days, it should be studied page by page and line upon line, by every historical writer, whether of the drama or of definitions in the dictionary, to which, indeed, it has contributed much valuable evidence.

I turn naturally to *Margaret* to find whether sun-dials were in common use in New England after the Revolution; here is a bit of a scene in an opening chapter, entitled, "Work and Beauty; an Impression of the Real,"—it is but a simple asking of the time-o'-the-day:—

"The child Margaret sits in the door of her house on a low stool with a small wheel, winding spools, 'quilling' for her mother, who, in a room near by, is mounted in a loom, weaving and smoking; the fumes of her pipe mingling with the whizz of the shuttle and the jarring of the lathe and the clattering of treadles. From a windle the thread is conducted to the quills, and buzz, buzz, goes Margaret's wheel, while a gray squirrel, squatted on her shoulder, inspects the operation with profound gravity.

"'Look up the chimney, child,' says the mother, 'and see what time it is.'

"'I don't know how,' replies Margaret.

"'I suppose we must get the Master to learn you your a b c's in this matter,' rejoined the mother. 'When the sun gets in one nick, it is ten o'clock; when it reaches the stone that bouges out there, it is dinner time. How many quills have you done?'

"'The basket is full, and the box besides. Chilion said I might go and sail with him.'

"'We have a great deal to do. Miss Gisborne's flannel is promised the last of the week, and it must be drawn in to-morrow. I want you to clean the skans; there is a bunch of lucks down cellar, bring them up; get some plantain and dandelion on the smooth for greens; you must pick over these beans, put some kindlers under the pot, then you may go.'"

Now! There is a half page of plainest description of the simplest home-life; yet almost every line, certainly every sentence, contains a word or phrase, or refers to a deed as obsolete and as absolutely incomprehensible to a New England country child to-day, as would be a sun-dial to him, or as was the time-marking of the open chimney-place to Margaret. I venture to assert, also, that half of my readers will possess a like ignorance. Nowhere throughout the book is a sun-dial referred to; and to me this proof is absolute of their rarity. If there had been a sun-dial, Margaret would certainly have run to it. Nor in the extraordinary Boston to which Margaret fled in her shadowed girlhood was there a sun-dial in the Wiswell garden; nor was there one in the wholly artificial garden and surprising home created for her as a wife.

One of the simplest devices by which the midday hour was made known to dwellers in rural homes earlier than Margaret's day was a noon-mark. The dweller in town or village had the noon bell from the church steeple, but on nearly every farm-house was a noon-mark, usually by a frequented door or window.

I have seen them many a time on the threshold of a barn, at the kitchen doorstep, or outside the pantry. Country folk grew very skilful in telling the relative time from a noon-mark. I knew one old woman who, by her kitchen noon-mark, could tell the hours from ten to four without a variation of four minutes, which is in general all that would be expected from a watch — from a woman's watch. Noon-marks have been set in the form of a line of colored pebbles in well-laid earth or cement at the base of some stationary pole or flagstaff. We have them in several of our "Homes" — or refuges for life-wrecked sailors and life-beaten soldiers.

To whatever country we wander we find among all uncivilized peoples this vertical pole fixed in the ground as a primitive gnomon. In India and other Asiatic lands the natives are wise in reading the hours of this simple dial, making it serve as an exact chronometer. The Labrador Indians when on the hunt stalk on in advance of the train with their arms; while the women, heavily laden with provisions and means of shelter drag along slowly after. When the lords and masters begin to think of food-time, or wish in any way to leave some guide as to their progress for the squaws, they thrust an upright stick or spear in the snow, and draw in the snow the exact line of the shadow then cast. The women, toiling painfully along, note the spear, and the progress of the shadow, and know closely the difference of time. They know, too, whether they dare to linger for a few minutes' rest, or if they must hastily catch stick or spear and wearily hurry on.

In Upper Egypt the hours for work on a water-wheel are still fixed by primitive sun-dials which are scarce more than noon-marks. One of these sun-dials is made by extending a maize or dhurra stalk north and south on two forked uprights. At the side are set in the earth pegs which evenly divide the space between the sunrise and sunset shadows of this dhurra stalk. In the other dial the gnomon is a vertical stick. Often the pegs are nearly covered by the soil, so firmly are they pressed in, in order to avoid being moved by the feet of cattle or men. The space between two pegs is called an *alka* from an Arabic root meaning to hang or hitch on. The harnessing of a bullock to a water-wheel is merely the hitching on of a loop of harness over a hook. To the question, What do you do when the shadow reaches this peg? the answer always came, "We hitch on another bullock." These sun-dials are constructed entirely upon observation, with no scientific knowledge. An English scientist was once asked by the celebrated Sheik Daig, as a test of his learning, to construct a sun-dial. While the Englishman was making full explanations of latitudes, horizontal planes, etc., the Sheik abruptly interrupted by thrusting his spear in the ground and marking therefrom on the ground the exact lines of shadow which would fall at certain hours of prayer. Though this primitive time-teller still is used, there are no ancient Egyptian sun-dials known; nor is it anywhere stated in ancient writings that the Egyptians used their obelisks as gnomons.

At Settle in Yorkshire, England, rises a hill

Natural Sun-dial at Settle, Yorkshire, England.

called Castleberg. Until about a hundred years ago a great mass of rock on that hill formed a natural sun-dial. It is shown rather crudely on this page in a reproduction of an old engraving, given in Smith's *Old Yorkshire.* It is thus described in the letters of Bishop Pococke, written in 1750, and now edited for the Camden Society:—

"Crossing the Ribble, we came in a quarter of a mile to Settle, a little town situated under a high rocky hill; on the lower part of which, four stones being placed, they serve as a sun-dial to the country for three or four miles southward, as they know what hour of the morn it is the shadow comes to them from nine to twelve."

The stones have disappeared, but the memory of the sun-dial still lingers as well as the drawings and engravings of it. Many "natural" sun-dials exist. In Ireland as late as 1813 very few persons owned clocks and watches, and each settlement had some natural dial by which the nine watches of the day were shown by the sun's passage over certain mountain-peaks, or over set pyramids of stones if the natural formations did not afford a suitable object as a marking place. In Norway similar modes have been used to mark the time. Simpler shadow-marks are known to many dwellers in isolated homes; and I well recall the deploration I heard some years ago in a New England village because a certain great pine tree which had cast a much-consulted noon-mark had fallen in a storm.

Since the year 1792 an obelisk has stood in the Piazza Monte Citorio at Rome. It has had a varied history, having been lost to sight for many years. Its entire service in casting a noon-mark, after it was first brought from Egypt, is thus told by old Pliny:—

"As for that Obelisk that stands as a gnomon in Mars Field, Augustus Cæsar devised a wonderfull means that it should serve to mark out the noontide, with the length of day and night according to the Shadowes which the Sun doth yeeld by it; for hee placed underneath at the foot of the said Obelisk, according to the bignes and height thereof, a pavement of broad stone, wherein a man might know the fixt hour at mid-day, when the shadow was equal to the Obelisk; and how little by little according to certain Rules (which are lines of brasse inlaid within the said stone) the days of increase or decrease."

Many noon-marks in foreign edifices are interesting; one at the church of St. Petronio, Bologna, is 220 feet in length and was traced in 1653. Another is at St. Sulpice in Paris. At Salisbury Cathedral a perpendicular meridian line has been marked on the north boundary wall of the Close. The spire of the cathedral serves as a gnomon and throws its shadow at noon. The word *Meredies* is engraved inside the line.

When the city of Albany had as a mayor a man of parts, Hon. John Boyd Thatcher, he caused to be placed in front of the City Hall a carefully traced brass noon-mark, or, more properly speaking, a meridian line. This serves to invite a vast number of inquiries and to elicit some surprising answers; one being that it "marked the end of a telephone"; another that it marked the boundary of an Indian grant; another from a boy who said in all seriousness that the mayor put it in for the boys to take "cat slides" on — a cat-slide being, I must explain to those who know not, a bit of clean ice on a city street whereon a boy — yes, and a man, too — can by a slight run have impetus to slide swiftly and happily on one or both feet to the very end. The brass meridian line serves so well this purpose, it cannot be wondered that the boy assumed it to be its only reason for being.

I would like to see these fine brass meridan lines much more frequent than we do on the floor of broad vestibules, of open porches, of large plazas, of paved terraces; wherever the clear sun rays can shine and prove the use of the noon-mark; and in one place I should like to see a noon-mark which would be of world-wide importance — at Washington.

I beg to call the attention of the Government of the United States and the Commissioners of the District of Columbia, to an opportunity of easily making the finest sun-dial in the entire world, or if not that, the finest noon-mark. The Washington Monument, with its superb shaft 555 feet in height, most glorious of all gnomons, traces unmarked day by day its wonderful parabolic path on the greensward around it. What a beautiful sight it would be if the Government would order the tracing of its analemma and mark the hours with beds of flowers! What an instructive and inspiring object it would be to all who visit that great Monument; there might arise from its inspiration some thoughtful youth, another Ferguson or Wren, to add to the list of the great mathematicians of the world. If a sun-dial is not traced, a meridan line positively should be set; a line of stone or white marble, a noon-mark in the grass. This would not equal the dial, but would be better than the unmarked round of to-day.

We had the meridian line in Washington surveyed and marked in noble fashion when the City and District were first laid out; and the most interesting meridian line in the whole world should naturally be to Americans this famous national meridian line of the United States; but it has fared at our hands as though it were an object of obloquy instead of pride.

On the 15th of April in 1791 there was laid with solemn and elaborate Masonic ceremonial the corner-stone of the District of Columbia; now half-forgotten and hidden from view, this mighty symbol of our vast nation forms part of the foundation wall of the lighthouse at Jones Point near Alexandria, Virginia. The ten miles of the District were marked during the following year with mile-stones, which bore numbers, and on the District side the words, "Jurisdiction of the United States," on the other the names of the surrounding states, dates, etc. These stones are known in their neighborhood as "Jurisdiction Stones." As it was then the custom of various great nations to reckon longitude from their own capitals,—and a bad system it was,—our Revolutionary ancestors promptly proposed that the new nation should have its meridian line. On L'Enfant's plan for the Federal city appears a mark for an historic column (now the site of the Emancipation Statue in Lincoln Park), and from this column all distances through the continent were to be calculated. But when Ellicott, another engineer, laid off

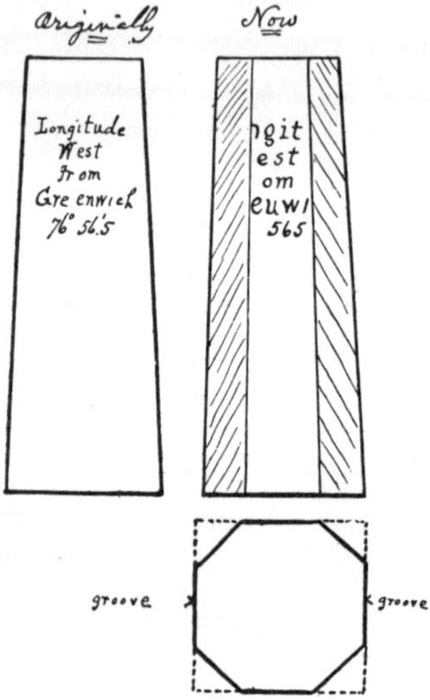

Drawing of the Meridian Stone of the United States of America.

the streets, avenues, and "appropriations," as they were called, he began by drawing a true meridian line through the "Congress House," or Capitol.

But the Meridian Hill known to old Washingtonians is not at the Capitol, but north of the White House, at the head of Sixteenth Street, so there is another meridian to consider. In a letter (now in the State Department) written to President Jefferson by Nicholas King, it appears that King laid out a meridian line along Sixteenth Street in 1804. The letter is given by Mr. Marcus Baker in his interesting article, "Surveys and Maps of the District of Columbia." An obelisk was planted on the top of a hill north of the president's house; and two stones were set near the site of the Washington Monument. This obelisk is gone and the site unmarked. Another was set near the Capitol and called the Capitol Stone; this is also vanished and the site unmarked. Another stone, known as the Jefferson Stone, was also set. The site of this is known.

Admiral Porter had a house at the head of Sixteenth Street; on the southern lawn stood a low sandstone block on which was placed a brass sundial. This has been called the Meridian Stone; it was removed and is now doing service as a carriage step at the corner of R and Fourteenth streets. This was not, so Mr. Baker infers, the original stone. The true meridian stone, set in 1804, stood where placed until some time in the seventies, when Meridian Hill was graded down. The stone was carried to the District building and thrown in a rubbish heap.

Later, when Lieutenant Hoxie was on duty, it was sent to the Reform School and set up as a hitching post. It is about four feet in height and was originally square in section and slightly tapering. It bore the inscription lettered thus:—

Longitude
West
from
Greenwich
76° 56′ 5″

The corners have been cut off, and this lettering only remains:—

igit
est
om
enwi
56′

There are many ways of making a noon-mark. A very unscientific but very satisfactory one is this: On either April 15, June 15, September 1, or December 24, the four days of the year when the sun and the clock are exactly together, secure a watch or clock, known to be exact by some standard time. Then on the surface where you desire to draw your noon-mark cast a straight shadow at twelve by your watch, and mark it definitely. Another way; is on any clear night, hang (out-of-doors) two plumb-lines in such a position that on sighting from one to the other the North Star will be in exact range. Drive two stakes exactly in the place of the two plumb-lines, and when the shadow at noon of one stake extends precisely to the other stake, that shadow-line makes an accurate noon-mark.

An interesting noon-mark has been for many years at Durham Cathedral, England, and is shown in a curious manner. About ten feet from the floor a thin piece of stone having in it a circular hole

Sun-dial at Elm Hirst, Wilmslow, England.

about an inch in diameter is inserted in a window. Through this opening shine the rays of the sun, throwing a bright spot of light, which at noon falls on the meridian line. This contrivance at Durham Cathedral forms one of a class called spot-dials, or when evidenced from reflected light, "reflective-dials." Such was the dial made by Sir Isaac Newton when a boy. He painted a dial-face on the ceiling of his room, and the spot of light was cast upon the hour lines by means of a bit of mirror fixed horizontally in the window ledge. This was in the house of his grandmother, Mrs. Ayscough. The plaster ceiling with the marks of the hours, etc., is still preserved in the new house which was built on the site of the house wherein this "ceiling-dial" or "reflective-dial" was originally made.

I own a number of old books on dialling, and I find these ceiling-dials a "favorite conceit" of the old diallers; one of them says: "I confesse it is a pleasant thing to behold how Art hath taught the Sunne to trace out those Lines and Parallels by Reflection from a Glasse, which his direct beames can never shine upon." In Leybourne's *Dialling* (my copy is a vast folio of the year 1700) many rules and designs are given. One of his window-dials I copy facing this page. In his rules many curious and antiquated terms appear, such as "quarrys" and "quarrels" of glass; the "jaums of a jetty window," the cheek-posts," etc.

By such rules as these was young Newton allured to try his skill. Leybourne's rules for making window-dials and reflective-dials are very clear and easy to understand.

An extraordinary ceiling-dial was made by Sir Christopher Wren when but a mere boy; it must be recalled that dialling was then a part, not only of an advanced education, but also of a plainer everyday schooling. Wren had translated, in 1647, Oughtred's *Geometrical Dialling* into Latin, when he was fourteen, and it had been published; and he had

Window-dial, Leybourne's *Dialling*.

thus acquired a knowledge both of mathematics to make this ceiling-dial, and of Latin for the inscription, which, translated, reads thus:—

"Chr. Wren: One who was content to depict upon this narrow ceiling the pattern of the sky, obtained from Phœbus the gift of a rival of his rays, an image upon a mirror; that would pass over this heaven with borrowed light and make an effigy of his yearly course. 1648 years after the time wherein in very truth Man was made God from a Virgin's womb, and in the 16th year of (the maker's) youthful age."

In this inscription the dates are formed by chronograms—the capitalization of certain letters in the last lines of the inscription; this was one of the fashionable fancies in inscriptions of that day. I have referred more fully to Wren and his interesting teacher in my chapter on "Ingeniose Diallers."

This ceiling-dial was but one of the "universally curious" works of what Evelyn called that "prodigious young scholar Mr Ch^r Wren." Evelyn saw at Oxford "a variety of shadows, dyals, prospective and many other artificial, mathematical, and magical curiosities, a way-wiser, a thermometer, a monstrous magnet, and other sections, a ballance on a demi-arch"—these the work of Wren and his teacher.

A way-wiser was an instrument known now as an odometer or perambulator—the Latin derivatives having replaced the simple old word, meaning a something to make you wise or knowing of the way you have fared. A way-wiser seems to have been for many years a sort of plaything of scientists and the scientific work of dilettantes. It has also been of practical use. Our own Franklin laid out our pre-Revolutionary post-roads with one attached to a comfortable chaise in which he rode, followed by carts bearing mile-stones. It has been more formally used in the preparation of our state maps and other important topographical work. In 1657 Evelyn saw Colonel Blount's way-wiser, which was attached to a coach, which "exactly measured the miles and showed them by an index as we went on." This way-wiser could measure up to one thousand miles. It was deemed a wonderful instrument and a rare one; but to-day along the roads so leisurely surveyed by Franklin, nearly every bicycle that flashes past his still-standing mile-stones bears a cyclometer — a modern and cheap way-wiser, beside which Colonel Blount's machine stands in the same relation as a sun-dial to a Waterbury watch.

A very interesting spot-dial was made by using a lens or sun-glass. In a garden in Cheshire, at Elm Hirst, Wilmslow (page 45), is a lens-dial on which is the appropriate motto: "WHATSOEVER DOTH MAKE MANIFEST IS LIGHT" (Ephesians v. 13).

Another use of a magnifying glass in a dial is shown in what are known as cannon-dials; these are found in several European towns. One is given on page 49 which was made for the Sultan of Morocco by Messrs F. Barker & Son of London. It is a beautiful instrument, being made of fine brass inlaid with white metal, and is an accurate timekeeper. In these cannon-dials the glass is so fixed that at exact noon the concentrated rays of the sun ignites the

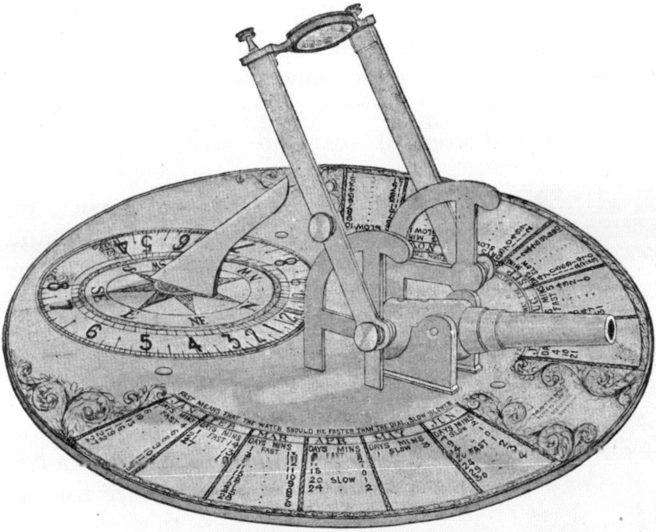

Cannon-dial of Sultan of Morocco.

powder in a touchhole and fires the cannon. Small sun-dials have been made after this pattern.

There is an interesting and unusual lens-dial at Frankford Arsenal near Philadelphia, which is mounted on the muzzle of an old iron cannon set vertically in that part of the arsenal grounds on which faces the government cartridge factory. It is shown on page 50. This ingenious lens dial was designed and placed in its present position by the late Captain William Prince, Ordnance Department U. S. Army, in the year 1874. The mounting is an unusually satisfactory one, for not only is it in good taste, being suited to the surroundings, but also of positive stability, warranting a perfect spirit-level for the dial-face, which is of much importance for the rather delicate contrivance which the dial displays for marking accurate time. It is protected against special stress of wear and weather by a hinged iron cap or cover.

Cannon-dial at Arsenal, Frankford, Pennsylvania.

The dial consists of a triangular gnomon mounted on a circular plate. The plate has Roman numerals for the hours, and exact tables of corrections to be made for true local time. The gnomon has a special feature for indicating the corrected time at noon; this is by means of a lens so mounted in its inclined edge as to project an image of the sun on to the annular surface of an opening through the gnomon. On this annular surface is inscribed a figure 8 loop of two equidistant lines between which the image of the sun appears at local noon in some part of the loop, varying with the time of the year. Unfortunately this contrivance does not show in the illustration. A similar arrangement may be found on a sun-dial at Monaco, where one is gravely told that it is "the only perfect sun-dial in the world."

There is something very suggestive of sentiment in the thought that in a spot-dial you tell the hour by a mark of light instead of shadow; and such a dial needs a special motto.

Several mottoes are given in Leadbetter's treatise called *Mechanick Dialling*, 1756. Their being in the same metre gives them the appearance of being machine poetry, written for or by Leadbetter for these spot-dials.

SEE THE LITTLE DAY-STAR MOVING
LIFE AND TIME ARE WORTH IMPROVING
  SEIZE THE MOMENTS WHILE THEY STAY
SEIZE AND USE THEM
LEST YOU LOSE THEM
  AND LAMENT THE WASTED DAY.

Another reads:—

SHINING SPOT FOREVER SHINING
BRIGHTEST HOURS HAVE NO ABIDING
  USE THY GOLDEN MOMENTS WELL
LIFE IS WASTING
DEATH IS HASTING
  DEATH CONSIGNS TO HEAVEN OR HELL.

In France a dial wherein the hour is shown by a ray of light is called *Cadran à La Capucine*. On such a dial in a Franciscan convent are these verses:—

Pourquoi sur ce cadran solaire
Ne voit-on point l'ombre ordinaire?
C'est que consacrant dans ce lieu
Tous notre temps à louer dieu.
Il faut pour le marquer lu plus noble manière
C'est d'emprunter au ciel un rayon de lumière.

WHY DO YOU NOT SEE THE USUAL SHADOW ON THIS SUN-DIAL?
IT IS BECAUSE IN THIS PLACE ALL OUR TIME IS CONSECRATED TO PRAISING GOD.
WE DESIRE TO MARK IT IN THE NOBLEST MANNER,
AND THAT IS BY BORROWING A RAY OF LIGHT FROM HEAVEN.

The shadow of the gnomon may be cast upon a window, and can thus be seen from within the house. This is called a refractive dialling, or a window-dial; and in thus viewing it from within doors, the shadow will appear to go round as do the hands of a clock, while in an ordinary vertical dial the reverse motion is seen.

These are sometimes of stained glass, and in England have been placed in churches. A fine one is shown on page 53, it is leaded into a window at Kersal Cell, near Manchester, England; the home of John Byrom, who wrote "Christians Awake!" I don't know why these nearly all have a fly painted on them — perhaps as a remote pun that the hours fly. The window-dial at Lambeth Palace, one at the private chapel at Berkeley Castle, both have the fly. Another has both a fly and a butterfly — the latter being the emblem of immortality.

The motto, *Dum spectas fugio*, is a favorite motto for these window-dials: WHILE THOU LOOKEST I FLY. Arthur Young, in his *Six Weeks' Tours*, tells of two window-dials at the Rectory, North-hill, Bedfordshire. He says that the fly had the wings painted on one side of the glass, and the body and legs on the other, so to deceive fully the spectator. The date was 1664.

The ancient Greeks wrote of measuring the day by the course of a shadow, and speak of a six-foot shadow, a ten-foot shadow. It has been suggested

Window-dial at Kersal Cell, Manchester, England.

that this was each man's own shadow as thrown on the ground; long in the morning and at night, and short at midday, and that he measured it with his own foot, as did the Malays in Madagascar.

The early successors of the noon-mark, such as the water-clock or clepsydra, were known to many nations in some form, though it is told that the clepsydra was invented 2636 B.C. by a Chinese scientist. Duke Chan, who is alleged to be the inventor of the compass about 1130 B.C., was the first to employ the clepsydra as a timepiece. Chinese poetry, even the most ancient, abounds in graceful and sentimental allusions to the clepsydra. A waiting wife complains of the leaden foot of time in the form of verse called a "stop-short":—

"It seems that the Clepsydra
Has been filled up with the Sea
To make the long, long nights appear
An endless time to me.

"The incense-stick is burnt to ash,
The water-clock is stilled,
The midnight breeze blows sharply by,
And all around is chilled."

Even by 1851 only one clepsydra was in official use; it was in the watch-tower of the city of Canton; my sister saw it there, still in use, in the year 1899. It consisted of four copper jars on a flight of steps, the top of each reaching to the bottom of the next in succession; small troughs connected them all. The largest jar held about ninety-three pints of water. A wooden index was set in the lower jar and rose as it filled with water. It was set at five in the morning and five in the afternoon. When the half-day was ended, the water from the lower jar was ladled back into the upper one by two watchmen, who also beat the twelve watches of the day on drums. The Chinese do not number the hours; they simply name these twelve divisions and designate each with a sign. This clepsydra is so rude a contrivance that it hardly seems fit for a race so civilized as the Chinese. These Cantonese folk would be far out in their time-reckoning if they depended on this ancient clepsydra and their time-sticks, which are sold by the man who has charge of this "copper-jar-dropper," as it is called. These are referred to in the second stop-short quoted above: "The incense-stick is burnt to ash." These time-sticks were made of sawdust (usually of a certain wood), a slight mixture of glue, rolled into even cylinders two feet long, and divided into hours. They consumed without flame, and burnt up in half a day. They are like the time-candles of other countries, and share the interest always inspired by every time-keeper. I remember well the fascination which King Alfred's "candle-clocks" had for me in my childhood; as told in a little book of anecdotes of English kings and princes. I recall well making candle-clocks from common wax candles, and our disappointment when they would not burn four hours precisely, as did the king's.

A burning candle was used in England and France in many special cases to mark a short extent of time; as an auction "by inch of candle," wherein the last bidder as the flame expired was the successful one. Servants also were bidden for and paupers "boarded out" by inch of candle.

The ancient clepsydra was sometimes extremely ornamental, the copper jars being made in the shape of dragons and other figures, and the index was also ornamented. Another clepsydra was shaped like a bird; the water fell from its beak and was received in a vessel on a balance. Another water-clock was a perforated copper vessel which was placed in a tub of water and filled gradually and sunk every hour. The Malays in their proas use a similar rude water-clock made of a perforated cocoanut shell. A more complicated machine which represented the motions of the heavenly bodies was run by falling water; it was a huge hollow globe perforated on its surface so as to afford, when lighted from within, a representation of the starry sky at night.

The *Liliwati*, a profound mathematical treatise of the twelfth century, was written by an Indian astronomer, who was "grievously baffled" of the marriage of his daughter named Liliwati. It was predicted that she should die unmarried; but the father determined to avert that disgrace. He found from astrologers a lucky hour, and secured a bridegroom. But the hour passed without being noted on the clepsydra, for a pearl from the girl's bridal dress fell into the bowl and closed the opening; and the bridegroom departed. The father consoled his daughter by writing this wonderful book which would transmit her name better than could any children. It is translated into English and published by a Calcutta firm, and is of great interest and research.

Clocks and watches are much cherished in China; ancient ones of very antiquated appearance are constantly seen in use; some of these are like the "Nuremburg eggs." As Chinese gentlemen carry two watches and are particular to have them harmonize, clock and watch menders find constant employment.

For their watch-making skill the Chinese are said to be indebted to the teachings of the Jesuit missions.

The European water-clock of the seventeenth century is described in Kirchner's *Ars Umbræ et Lucis*; and in the form there presented is almost universally attributed to the Jesuits. Pewter clepsydras were made in considerable numbers in France.

I am informed that a picturesque water-clock or "hour-bowl," shaped like the Chinese water-bowl, is still found in remote parts of India; picturesque as absolutely simple things can be, and generally are. A globular copper bottle or bowl has a hole in the bottom. The water runs slowly through the little orifice until the bowl is empty, when a waiting attendant strikes the empty vessel a resounding blow with a hammer; then he refills it, and hangs it up to drip again. Of course this has to be made of an exact size proper to measure an hour.

It is told that in some Oriental countries a stone is flung in the bowl and thus resounding strikes the hour. The opening stanza of Fitzgerald's translation of *Omar Khayyam* in the original edition ran thus: —

> "Awake, for Morning in the bowl of Night
> Has flung the stone that puts the Stars to flight;
> And lo! the hunter of the East has caught
> The Sultan's turret in a noose of light."

I should be convinced that these stirring lines referred to an emblematic use of the ancient Oriental time-bowl, save for one thing: they were not written by the old Persian at all, but were wholly Fitzgerald's thought and words; and help to prove, what we all know, that this is a case when the translation is far greater than the original. Sand-glasses, or hour-glasses, were first made about the year 330 A.D. There are other dials of the ancients which fascinate the antiquary,—wind-dials or anemoscopes,—in which the courses of the winds were marked on a dial connected with a weather-vane. They have been found in Pompeii and in Rome. The most famous was known as The Tower of the Winds, an octagonal horologium which was one of the wonders and beauties of ancient Athens. It is pictured opposite this page. The bronze Triton which served as a weather-vane has vanished, but eight sculptures remain. These bold flying figures represent the winds, and under each was once a sun-dial. There was also a water-clock. As the tower was forty feet in height and twenty-seven in diameter, it formed a striking object. Boreas, the North wind, blew on a conch-shell; the South wind poured rain from a water-jar; Zephyrus carried a mantle filled with flowers.

This Tower of the Winds is the oldest known construction for observing the winds, but a similar pillar covered with copper was at Constantinople; both of these towers had weather-vanes. For a time it would seem that only important buildings, chiefly churches, carried vanes. In France in the twelfth century none but noblemen could have weather-vanes, and for a time no noblemen save those who had planted their standards on some rampart at the storming of a town or citadel. These vanes then bore the knight's arms. On the Bayeux Tapestry

Tower of the Winds, Athens.

ships appear, and these have vanes on the masts. Anemoscopes, to show the duration of the wind, and anemometers, to measure its force, have been invented in many shapes; one resembled a wind-mill. Both instruments were in use in England in Queen Anne's time. They were fixed in coffee-houses where merchants and ship-owners congregated, and where winds and weather formed a constant and natural topic of conversation.

It is probable that clocks may have been regarded with suspicious eye by the distrustful and superstitious pedants of the day when they were first made. Everything unusual, and above all everything clever, was adjudged to be akin to witchcraft — until it was proved not to be. The very first naming of a clock (so-asserted), in 1449, is by one Dr. Peacock, Bishop of Chichester, and he says: —

> "In all Holie Scripture it is not expressid by bidding counseiling or witnessing or by any ensaumbling of persoon . . . that men schulde mak and vse clockis forto knowe the houris of the dai and nygt, for thow in Scripture mensionn is maad of orologis schewing the houris of the dai by schadow maad by the sunne in a circle; certes nevere saue in late daies was any clok tellyng the houris of the dai and nyghte by peise and by stroke," etc., etc.

I suppose there were old fogies in that century as ever since, who declared that the clocks were a nuisance, that they were kept awake by the striking; and that the Evil One must have had his hand in them; that they were an unnecessary expense, being naturally, in the beginning, a constant outlay for repairs; that they would encourage the young folks sitting up late, would waste candles and fire, that the good old ways and the good old sun-dials and noon-marks were good enough for *them*, and ought to be for their children. Clock or Automobile! it doesn't matter much which; it is only a difference in dates and as regarded in comparison with other things.

CHAPTER III

CLASSIFICATION OF SUN-DIALS

"In which very many sorts of Dialls are Conteined, by which besides the houres of all kinds diversely express'd. Amongst which very many Dialls, especially the Most Curious are new Inventions hitherto Divulg'd to none. All these Particulars are Shortly yett Clearly sett forth for the common good."

— *Title page of The Explication of The Diall Set up in the King's Garden, An. 1669.* FATHER FRANCIS HALL.

OF DIALS we may form two classes, portable and fixed. Portable dials have in this book a special chapter; as do also noon-marks or meridian lines and spot-dials, ceiling-dials, and window-dials which are not in one sense separate objects. Other dials will be classed according to the divisions of Messrs. Ross and McGibbon, as formulated in their volume on Scottish sun-dials in their book of several volumes, entitled *The Castellated and Domestic Architecture of Scotland*. The sun-dials of Scotland are the finest in the world, and this noble book is worthy its subject. About two hundred Scottish sun-dials are described in it, and thus a clear idea is given of the art of dial-making of the sixteenth to the nineteenth century, as practised in Scotland. I am glad to adopt in general the simple and lucid classification of sun-dials as arranged by these two skilled and learned architects, and to use somewhat their words in describing the classes. The first division is into two great classes, the attached and the detached; the former being, as the name would imply, displayed upon the walls of some building, and often of slight importance in the general scheme of the edifice; while the second class, dials standing on their own special pedestals, are often of much monumental importance. The attached dials are divided into six classes: —

1. Single-faced dials.
2. Dials with two faces placed generally on corners of buildings.
3. Dials with two or more faces projected on corbels.
4. Terminal dials.
5. Dials on market and other crosses.
6. Horizontal attached dial.

In the first class the dials of a single face, of stone, wood, or metal, are usually affixed to the surface of a building, or they may be carved or painted on the wall itself; but sometimes, when the wall of the house does not face exactly as desired, these are canted out a bit from the house at one edge. That at Fountainhall, East Lothian, is thus placed at a slight angle that the dial may face due south.

Examples of these single wall dials are many throughout this book: they are a favorite dial for church decoration and use, and are found on many of the cathedrals and churches of the old and the new world, where they form the very best and most appropriate ornamentation a church can have ; even

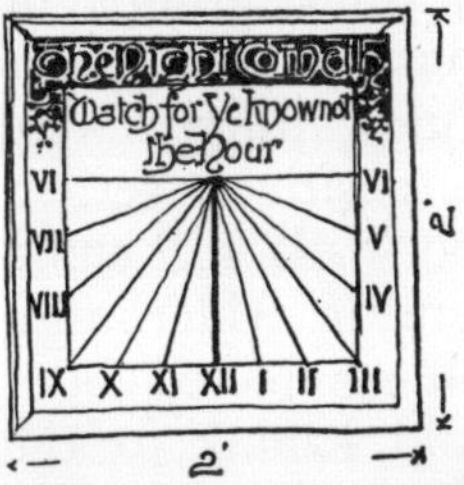

Grace Church, and Dial, Merchantsville, N. J.

a simple modern one such as this on Grace Church, Merchantsville, New Jersey, adds much to the beauty of a church edifice. A vertical dial is, in one respect, much better fitted for a public building than a horizontal dial, for the latter through careless or mischievous handling often becomes so shaken as to be useless as a timekeeper. A fine horizontal dial, with an unusual and beautiful gnomon, shaped like an angel's wing, was set but a few years ago as a memorial of a son to his mother, in the churchyard of All Saints' Church, Worcester, Massachusetts. Even this firmly set and carefully made bronze dial-face has been pulled and twisted by boyish visitors to the racking of the gnomon. A vertical wall-dial would have been out of mischief.

On those interesting old buildings — moot-halls — are found some of the most ancient dials. A moot-hall was a building with a room for public debate or a court, a town-hall. In early English history we find that a moot or meeting was constantly being held, and a court formed which had special powers. A town-moot was a meeting of the town; there was also the hundred-moot, a meeting of the hundred; and a folk-moot, a muster of all the people far and wide. The form of holding the court was the same in all; the priests proclaimed silence, appointed speakers on either side of the controversy who told their tales, while groups of freemen assented or dissented, shaking their spears and clanking their shields; sometimes witnesses were called; and finally everything was positively settled by shouts of Aye and Nay. Moot-halls still stand in ancient and stagnant towns; one in Alderburgh is shown on page 10 — a lonely building erected about the year 1500. The dial is not so old, I think, for the motto, *Horas non numero nisi serenas*, seems hardly suited to a moot-hall of that year. It has a second motto also, *Pereunt et imputantur*.

A favorite position for vertical church-dials is over a door. The fine dial shown facing this page

Sun-dial over Church Porch at Eyam,
Derbyshire, England.

is over the south porch of the church at Eyam, Derbyshire. It has upon it the lines of the Tropic of Capricorn, the Tropic of Cancer, and the Equinoctial plainly marked and lettered; also the names of various cities and places,— Mexico, Panama, Quebec, Bermuda, London, Rome, Constantinople, Mecca, Calcutta, and Surat, showing their difference of time; also the mottoes *Induce Animum Sapientem* — TAKE TO THYSELF A WISE WORD; and *Ut Umbra sic Vita;* the date 1775, and the names, Wm. Lee, Thomas Fraggatt, Church Wardens. Surrounded by ancient elms and sycamores stands this venerable church, celebrated both for the singular Runic cross in its yard, and for the pathetic story of its heroic rector, Mompesson. In July, 1665, the plague was brought to Eyam in a box of tailor's cloths and patterns. It spread immediately and every house soon had its victims. Mompesson inspired the villagers with a wonderful spirit of self-abnegation, so that they cut themselves off from the rest of the world, in order to stop the spread of the pestilence. Not a soul left the place, and supplies were brought in from outposts. Pest-houses were established, but even winter did not subdue the disease; and in the spring Mompesson, knowing well the danger of any assembling within the church, preached to his heroic people from a projecting rock in the cleft in the dale, — a rock still known as "The Pulpit." Out of a village of three hundred and fifty inhabitants but ninety survived. His own wife died in August, and is buried in the churchyard. Many others were buried in the fields, and one little enclosure is shown which holds seven of one family, who died in seven days.

A very interesting and ancient sun-dial is on the south porch of Lewannick Church, and with the Cresset stone standing on the left of the porch, is deemed somewhat of a curiosity. This church was

Sun-dial on Chimney, Petersfield.

about five miles from Lanceston, but has just been destroyed by fire.

Sun-dials were placed on many symbolical or significant structures. Among these is the curious old triangular Lodge at Rushton, Northamptonshire, a monument of the deep Trinitarian belief of Sir Thomas Tresham, a skilled architect of Elizabeth's day. Its walls are covered with pious emblems and inscriptions. This lodge has three walls, three gables on each wall. On the central gable of each is a sun-dial; on the first is the word *Respicite;* on the

Cresset Stone and Dial on Lewannick Church, Lanceston, England.

second, *Non Mihi*; on the third, *Sole Laboravi*, 1593. "MARK YE, NOT FOR MYSELF ALONE HAVE I TOILED." The trefoil also is expressive.

An appropriate and well-designed modern setting for vertical dials upon dwelling-houses, and one I always like if not placed too high, is upon an external chimney. On page 66 is given a dial at Petersfield thus placed, and it may plainly be seen how well it suits its position.

On this page is shown the vertical sun-dial at Oak Woods in Peace Dale, Rhode Island, once the home of the old philosopher, Rowland G. Hazard, and now the home of his grandchildren. Another fine vertical dial, upon page 68, was set in 1891 on the old stone wall which marked the northern boundary of the grounds of the Santa Barbara Mission, California, in a place where many passers-by see and use it. The inscription reads:—

Wall-dial at Oak Woods, Peace Dale, Rhode Island; residence of Rowland G. Hazard, Esq.

"The dial was made, inscribed, and set by Rowland Hazard of Peace Dale, Rhode Island, in a part of the Sta. Barbara Mission Wall, built 1786, standing on his land."

The words *Fidem docet* of the noble motto are illumined with sunlight, as if prophetic.

Lux dei vitæ viam monstrat
Sed umbra horam atque fidem docet.

THE LIGHT OF GOD SHOWETH THE WAY OF LIFE,
BUT THE SHADOW BOTH TELLETH THE HOUR AND TEACHETH THE FAITH.

The picture speaks to me many words besides those of its motto. It makes me think of the aged philosopher, Rowland G. Hazard, the most

Sun-dial at Santa Barbara Mission, California.

interesting person I ever knew, the friend of other philosophers, such as John Stuart Mill, and the author of such profound books as *Man a Creative First Cause*, *Freedom of Mind in Willing*; and I think of his sturdy race, some of whom were giants in the land; I recall the noble life of his son, who set up the dial; and I think of the philosopher's granddaughter, the president of Wellesley College, and her great work. Then I look at the picture, and the tropical foliage of the Cactus plant symbolizes to me folk of vastly different type—the native Mexicans and Californians, halting in the welcome shadow of the old wall, and reading with ease the Latin verses in their softened Spanish accent. I see the many Chinese wanderers to that shore, chattering the time of the day—for the sun's shadow speaks in their tongue as well as in Latin. I see Father Junipero and his barefooted and cowled Franciscan friars patiently teaching their trying converts—the Indians. I see behind the wall the Mission garden with its wells and cisterns, its dense Pepper trees, its Daturas and Roses; and Latin and Spanish, Asiatic, Indian, and Yankee races all blend in the spirit of this useful, beautiful sun-dial.

On page 70 is given an historical example of the second class, dials with two faces, set at the corner of houses. These are on the house in Edinburgh known as John Knox's house. On the corner of the house is carved a figure of the prophet Moses, kneeling and pointing with his right hand to a figure overhead,—namely, the sun in glory, on which is carved the name of God in three tongues. Beneath this figure were two sun-dials supported on iron brackets, which represented flames of fire. A good

John Knox's House, Edinburgh, Scotland.

example of these corner dials is shown facing page 72, a view of Mr. Thursby's house at Lower Harlestone, Northamptonshire. This is a charming English home picture; the cheerful house, the creeping vines and flowers, the tame pea-hen, and the sun-dials,—why, even the printed words *peacocks* and *sun-dials* give a picture of English country life! The house is about a hundred years old, and in 1891 the dials were nearly obliterated and the gnomons gone, but the dials have now been restored. The south-east dial is inscribed, THE HOUR IS AT HAND; the southwest, WATCH AND PRAY.

I have never seen in America a vertical dial of the third class,—namely, on a corbel, though the latter forms a most effective piece of architectural decoration as well as a useful one.

On Heriot's hospital in Edinburgh are eleven corbel dials, the finest attached dials of this type in Scotland; they are supported on carved brackets. Some of these brackets are cupids' heads with wings, others demons' heads and wings, one an elephant's head.

Terminal dials, the fourth class, form the apex of a gable, buttress, or some other portion of an edifice. They were a favorite decoration in Scotland; thus the church at Costorphine, Midlothian, has seven of these dials forming the finials of buttresses. On the eaves of gables, even of humble cottages, they were much used, and over the lichgates of churches formed a most appropriate finial.

The fifth-class dials on market and other crosses might well have been transferred to the class of detached dials, since the cross is often scarce more than a pillar to hold the dial. Many of the early market crosses and churchyard crosses of the seventeenth century bore sun-dials, and the earliest English sun-dial is of this type. These crosses varied from the simple village cross bearing a block with dials to the superbly ornamental market crosses of many Scotch and English towns, and many good examples will be shown in the pages of this book.

Near the Scotch border, in Cumberland, stands the little town of Bewcastle. It is in what has been ever a rather wild and desolate district, though it is supposed to have been a Roman station during the building of the famous wall. A view of the churchyard is given facing page 32. I think it the loneliest scene that I ever beheld—in which there are ample signs of both life and death. In it may be seen what is one of the most interesting ancient monuments in Great Britain; it affords the earliest English sepulchral inscription, the earliest piece of English literature, and of special note to us, the earliest English sun-dial.

It is a beautiful monument, too, a monolith fourteen and a half feet high, originally the shaft of a fine cross which added two and a half feet to its height. The cross blew down about three hundred years ago, and was sent to a collector of antiquities, Lord William Howard, for preservation; but its whereabouts is now unknown. The carvings on this shaft are very fine and pleasing, and, curious indeed, are in close relation to Byzantine art. It bears many runes, and on the south face is the sun-dial surrounded by carved bands and leaves, in such

Corner-dials at Lower Harleston, Northamptonshire. Residence of Mr. Thursby. *Circa* 1800.

a manner that the dial is proved part of the original design, not a later addition. The date of this sun-dial is 670 A.D., for the inscriptions have been deciphered and translated thus by the Bishop of Bristol: —

> "This thin sign of victory Hwaetred Wothgar Olwfwolthu set up after Alchfrith once King and son of Oswy. Pray for the high sin of his soul."

The date is given as the first year of the reign of King Ecgfrith. Other names appear on the shaft, — names of princes who served the cause of English Christianity in the seventh century. Oswy and Alchfrith were also Christians.

The sun-dial, which is within a third of the top of the shaft, is divided into four spaces — according to the octaval system of time measurement of the Angles. These four spaces are divided each into three spaces, making thus the twelve divisions of the day of the Romans. The gnomon was doubtless set horizontally, and could not have afforded a very acurate time-keeping, save for the noon-hour.

The loneliness of Bewcastle has aided in the preservation of this beautiful monument, and many simple ancient customs also lingered long. Among them was the *whittle-gate*, part of the payment of both schoolmaster and parson, and by which they found a home. The master lived in turn for two or three weeks at a time at the home of each of his scholars. As the houses were often scantily furnished, he carried his own knife or *whittle* for table use. Sometimes the parish gave the parson his whittle. We brought to America the custom of *whittle-gate*; it has lingered to the present day in remote country communities where the schoolmaster "boards around"; and we brought the word *whittle*, too, and applied it in such fashion that it serves to indicate a significant characteristic of native New Englanders.

Opposite this page is shown one of the three ancient dials of this class which are standing at Elmley, Worcestershire. One is at the meeting of two roads, a cubical dial on the shaft of an old cross. The second is in the churchyard at Elmley Castle; it is a cube with many hollowed-out dials, in some of which the gnomons remain. The third, here shown, is erected on the base of an old cross; the foundation is of several courses of masonry rising nearly three feet, and above them a stone so much like dial number two that antiquaries believe they originally formed a single dial. Besides the singular scooped-out depressions, this has also a shield bearing the arms of Savage, borne by the lord of the manor, which was granted by Henry VIII. This ancient block is surmounted by another, more modern, bearing four vertical dials.

The sixth class, horizontal attached dials, such as those fastened on window-sills or the parapets of bridges, are comparatively few in number and of no great distinction. I have several, captured from old window-ledges in New England.

Detached dials are divided into four classes: —

1. Obelisk-shaped dials.
2. Lectern-shaped dials.
3. Facet-headed dials.
4. Horizontal dials.

Antique Sun-dial in Churchyard, Elmley Castle, Worcestershire, England.

Naturally an exact line cannot be drawn between these to separate wholly each class; for a lectern-headed dial may have a pedestal somewhat obelisk-shaped, and the horizontal dial often runs into the other shapes; but the classification is as explicit as is possible.

The term obelisk-shaped is perhaps as good a descriptive word for the first class of dials as any single word could be; though the word obelisk conveys in general the thought of a plain shaft like the Egyptian obelisks. Obelisk-shape in a sun-dial, however, indicates a square shaft, supporting a bulging capital, and that surmounted by a tapering finial. Let me illustrate by referring to page 6 as an example; this is the sun-dial at Kelburne House, Ayrshire, and has all three characteristics. The shaft and the capital are divided into compartments which are hollowed out with "sinkings," which may be triangular, star-shaped, club-shaped, shell-shaped, shield-shaped, cross-shaped, liver-shaped, lozenge-shaped, circular, heart-shaped, and other geometrical figures. In these the sharp edge of the figure casts a shadow on the sunken cups; or a metal gnomon might be set thereon. Sometimes the hour lines are finely delineated. The bulging capital should be octagonal and have dial-faces on each of the eight sides, which dials might be sunken; and it could have also reclining and proclining dials (which should not be sunken), as has this one of Kelburne House. It will be noted that where the square shaft meets the octagonal pillar the triangular pieces are cut out, giving effective shadows and odd outlines. The tapering finial may have dial-faces but never sunken ones. The north side of the dial was usually left free for dates, initials, and inscriptions. This Kelburne dial is surmounted by a wrought-iron vane which encloses the monogram of the Earl of Glasgow and his wife; and it is dated 1707. The entire height of this is eight feet six and one-half inches without the steps at the base. These obelisk-shaped dials are usually set upon a platform or pedestal of steps, often of a circular or octagonal form; but a similar dial at Kelburne House is set in a basin of water, like a fountain basin — an unusual and inappropriate placing.

It will be known that when these dials were made of soft sandstone, as were so many in Scotland and England, the various shaped sinkings would quickly be worn so that they were of slight value as time-keepers, and the great number of dial-faces was confusing; so I ever regard the obelisk-shaped dial as a thing of beauty and dignity rather than as a precise timekeeper. Modern dials for fine gardens have often been made partly in the obelisk-shape; there is a beautiful one of these at Linburn House, which is shown on a later page.

The descriptive name of lectern-dial has been given to a sun-dial set upon a shaft which was never cut into dial-faces, but had a dial-head cut in a peculiar shape, so it resembled a music-stand or lectern. They were comparatively common in England and Scotland, though they were in general very complicated in their cutting, having several dial-faces, and being hollowed out, bevelled, and shaped in ways exceedingly difficult to describe and often most varied in each sun-dial. As an attempt at description, it may be said that a block of stone rests on the top of the standard somewhat as a book rests on a lectern; this block (the dial-stone) is cut into a shape somewhat resembling a Greek cross with semicircular depressions carved out of the four arms of the cross in such a way as to leave eight points or horns. Let this queerly cut cross be set well up on the face of the lectern; and have the circular depression at the top edge of the cross continued down the sloping back of the dial in a semicylindrical, trough-shaped opening. The shape is so singular, so complicated, and, to a careless observer, so unnecessarily complex and unnatural, that many speculations have arisen as to the reason for these forms. They were called Masonic dials and may have had some Masonic significance. An ingenious suggestion of ex-

Lectern-headed Dial at Gray House, near Dundee, Scotland.

planation is given through an article in the *Magazine of Art*, November, 1891, entitled "The Mystery of Holbein's Ambassadors." In this picture, part of which is reproduced opposite this page, is a table upon which are displayed several sun-dials and curious astronomical instruments. One of the last-named was "The Torquetum of Apian," by means of which the position of the sun, moon, or stars could be indicated at any hour. It is shaped like these lectern-dials and may have afforded the model for them. Though this form seems so forced and so contorted for its purpose as a dial, it was nevertheless used for many costly dials, which well illustrate the magnificent ideas of the seventeenth century in regard to the architectural accessories and furnishings of gardens, pleasure grounds, and parks of that date. The superb dials of Dundas Castle, of Neidpath Castle, of Mid Calder House, of Ladyland's House, show types of these costly dials; and the most elaborate of all is at Woodhouselee, Midlothian, carved exquisitely in the hollows of the shaft with the Thistle and the Rose. On page 77 is given a drawing of the sun-dial at Gray House, near Dundee. In this the ancient lectern-head has been set on modern steps.

In a general way the cross-dials seen in churchyards and graveyards may be deemed of the class of lectern-dials; though they bear slight resemblance to the elaborate Scotch dials.

The third class, that of facet-headed dials, may be described as consisting of a pedestal, baluster-shaped, bearing a spherical-shaped stone cut in a variety of

Instruments and Figure from Holbein's Ambassadors.

facets on which sun-dials are formed, cup-shaped, heart-shaped, or flat. This sphere or head is usually attached to the pedestal by a small pivot or bar.

Queen Mary's Dial, Holyrood Castle.

This shape perhaps affords to us the most beautiful of ancient and modern dials. One of the most interesting of all facet-headed dials, and one from which many others have been shaped, is called

Queen Mary's Dial at Holyrood Castle, Scotland, see page 79. It is said to have been the gift of Charles I to his Queen Henrietta Maria, and was a costly token; for in the accounts of the Master of Works it is shown that the sum of £408, 15*s.* 6*d.* was paid to the mason and his servants for "hewing of the diyell," and £66 13*s.* 5*d.* to a limner for gilding and engraving the dial. The dial stands with its base ten feet three inches high, and after lying for a long time in an apparently ruined condition, was repaired and reset by order of Queen Victoria. The facets of the dial-head have sinkings of many shapes,—heart-shaped, cup-shaped, triangular, square,—and the gnomons have many forms, one a grotesque face profile, another a Thistle. The dial also bears the Royal Arms and the initials of Charles and his queen. It has been copied, with some alterations and adaptations, in many garden dials, several of which are shown in these pages.

A superb dial of the facet-headed type is at Mount Melville, Fifeshire. It rivals the wonderful dial at Glamis Castle. A beautiful photograph of it has been given me by Miss Balfour-Melville, and is shown in this book. There are in all seventy dials on this grand structure, twenty-five of which are on the facet-head. This head is about eighteen inches in diameter, set without a pivot on a block beneath which is eleven inches square; this has three cup-shaped dials. On the shaft are rows of dials: plain dials, oblong-shaped sunken dials, heart-shaped dials, cup-hollows, and cylinder-shaped hollows.

A block of stone in the form known as an icosa-

Sun-dial at Mount Melville, Scotland.

hedron, having twenty faces, each being an equilateral triangle, is sometimes seen with a dial on each face. This facet-headed block is set on a pillar and forms a wonderful ornament for the garden.

Horizontal dials are subdivided into two classes, but I deem such division superfluous. It is said by Messrs. McGibbon and Ross that horizontal dials on pedestals are so numerous in Scotland that a list of them would include the name of every parish in Scotland; they must number hundreds. As no two are precisely alike, they would form a fine series of examples. I have studied the pedestals with great interest, since nearly all the garden-dials set up in America are of this class.

There are, of course, many forms of dials of which I have not spoken, among them the earliest dials in England; on old churches mysterious little rayed circles like the face of a sun-dial appear, which are often found in most out-of-the-way places, high up on the wall or underneath shelves. These are commonly called Saxon dials, and they strongly resemble "sun-circles" found on pre-historic remains, and are held by many scientists to be either sun-circles or luck-circles. There is usually a little hole in the centre of the rays, but of too shallow incision to hold a gnomon. The rays vary in number; some dials had sixteen. At Aldborough is a beautiful dial carved with the swastika or fylfot, resembling the sun-wheels of Danish relics of the Bronze Age.

It is said that there is hardly an unrestored church in Warwickshire, Northamptonshire, or Leicestershire but has these circles or imitation dials on its walls. On this page is given one of these rayed circles, which is on the porch of the Church of the Holy Sepulchre, Northampton. Its lines are most distinct. In an interesting paper called *Squints and Dials*, Sir Henry Dryden says that those

Saxon Dial on Church of the Holy Sepulchre, Northampton.

"Saxon dials" which were cut only breast high or where the central hole could not have held a gnomon, or where the rays are irregular, must be regarded with suspicion. I may define a squint for two reasons: first, for the information of American readers to whom the term is generally unfamiliar; and second, because they have so often been carefully pointed out to travellers in England as sun-dials. A squint is an oblique opening through the wall of an old church, cut in such a manner and for the express purpose of permitting a person in a transept

Model of Proclining Dial, Jeypore.

or side aisle to see the high altar at the time of the elevation of the host. The ordinary position of a squint is on either side of the chancel arch; but wherever they are, they always open to give a view of the altar. They are in general about three feet high and two feet wide, but at Minster Lovell,

Oxfordshire, they form narrow arches twelve feet in height. The word *hagioscope* has also been given them, also *squinch* — which is incorrect.

Vitruvius Pollio Marcus, a famous Roman architect and engineer under Cæsar and Augustus, wrote a treatise on architecture,—the only surviving Roman treatise on the subject. In it he names with very meagre and sapless description thirteen different kinds of sun-dials and gives the names of their inventors. During the careful archæological excavations of the past century many of these types have been found. I shall not attempt to explain or describe the thirteen types nor give illustrations of the hemispherum and hemicycle which were the earliest forms. These are never made to-day save in what our grandfathers would term " a rare conceit." In the Leyden Museum, the British Museum, the Louvre, examples may be studied. They are in general imperfect and very simple in original shape and decoration, except in one or two cases.

Those who wish to learn of the antique dials of Greece and Rome and of other rare forms of English and Continental dials should read Mrs. Gatty's *Book of Sun-dials*, or its enlargment by Mrs. Eden. " How I love those large still books," said Tennyson of one of Richardson's novels. Such a book is Mrs. Gatty's *Book of Sun-dials*. Like the sun-dial itself it seems associated with refined and serene things: the ivy-grown wall of the village church, the solemn graveyard, the ancient market square, the Rose walk, and Yew hedge, all of a day removed from present bustle and rush.

The most celebrated antique dials are those on the Tower of the Winds, referred to in the preceding chapter. It is impossible to give a date to these dials, but archæologists suggest that they are what are called quiver-dials by Vitruvius, wherein the lines, diverging from a common centre, resemble somewhat a sheaf of arrows tossed from a quiver. A

Quiver-dial of Phaidros.

dial of similar type is depicted here, the fine marble dial of Phaidros, which was brought from Athens by Lord Elgin, and is now in the Inscription Room in the British Museum. It bears a Greek inscription meaning, " Phaidros, son of Zoilos, a Pænian, made this." Phaidros was an architect who lived in the second or third century A.D.

Varied and curious in shape are the dials of Oriental lands. Sun-dials are placed on all the principal mosques in Constantinople. On many of them,

Model of Reclining Dial, Jeypore.

in addition to the lines necessary to indicate the course of the sun, there is a line drawn pointing to Mecca, whither the faces of the faithful must be turned while praying.

The superb dials of Delhi and Jeypore, constructed in 1724 by Rajah Jey Singh, are beyond description. They were built of vast size, of solid masonry and marble. The length of the hypothenuse of the gnomon was one hundred and eighteen feet. Photographs of the models of the sun-dials in the Victoria and Albert Museum, South Kensington, have been taken specially for this book.

Model of Sun-dial at Jeypore, India.

CHAPTER IV

THE CONSTRUCTION OF SUN-DIALS

"Here have wee then the Art of Numbring TIME by SHADOWES after the most *Methodicall Compendious* and *Perspicuous* Manner *compleatly* and demonstratively delivered for all Planes both by *Lines* and *Numbers*. The WORTH of this Work will be best valued by those who after sad wanderings have at length sate down wearied, with the obscure and toylsome Labyrinths of others."

—*Sciographia, or The Art of Shadowes*, JOHN WELLS, ESQUIRE, 1635.

QUOTING from a mathematical treatise, we define dialling or gnomonics as treating of the construction of any instrument, portable or fixed, which determines the divisions of the day by the motion of a shadow of some object on which the sun's rays fall.

A short definition of different mechanical forms of sun-dials, and of a few of the terms used in dialling, will be necessary for the full understanding of this chapter, and of any rules for the construction of dials. Dials are portable and fixed. The former are described in another chapter. The term fixed dial is applied to any dial fastened in a permanent position.

The dial may be horizontal or vertical, or inclining, which latter means fixed in any sloping position, fixed on any place that is not horizontal or vertical. A declining dial is one which does not face exactly either north, south, east, or west. A proclining dial is one whose plane is not vertical, but leans forward at the top; a reclining dial is the opposite, where its plane slopes backward. A refractive dial uses refracted light; such are the window-dials described in another chapter. A reflective dial is one in which the time is indicated by a spot of light thrown by a bit of mirror; such are spot-dials. A cylindrical dial is one drawn on a cylindrical surface, usually a half-cylinder; these are found on lectern-shaped dials. A globe-dial, star-dial, and a cross-dial are in the forms indicated by their names.

A quadrantal dial is in the shape of a quadrant, usually in portable form. An equinoctial dial is one whose plane is perpendicular to the earth's axis. The expression to rectify a sun-dial means simply to prepare a dial for an observation; to orient a dial is to place it properly in regard to the points of the compass.

A nocturnal dial shows the time by the moon's shadow, or by some mechanical device. A very quaint nocturnal dial from Leybourne's *Dialling* is here reproduced.

For the Centennial Exposition in Philadelphia, in 1876, there was made and set up an interesting sun-dial, having in all sixteen dial-faces. Among them were many of the unusual dials just named. This dial is an instrument of great scientific value, and

Nocturnal Dial. From Leybourne's *Dialling*.

has been preserved through the thought and care of the late Commander Joseph Philip Mickley, who had it placed upon the lawn of his home near Allentown, Pennsylvania, and carefully adjusted and restored. Mrs. Mickley has given me two photographs of this interesting dial; they are shown on pages 90 and 92. The dial gives the time at many of the principal cities of this globe.

The elementary astronomical facts upon which dialling is founded are the two motions of the earth, diurnal and annual. The correct way of expressing the two motions is, of course, that the earth turns on its axis uniformly from west to east in twenty-four hours, and is carried around the sun in one year at a nearly uniform rate; but in dialling the explanations are easier if the idea of the ancients is adopted,—the apparent motion, which is that the sun and stars revolve around the earth's axis once a day, the sun lagging a little behind the stars until its day is four minutes longer, and then back again. This retardation of the sun makes the time as measured by a sun-dial somewhat irregular, and a dial or clock which marks uniform time agrees exactly with a sun-dial but four days a year. These are April 15, June 15, September 1, and December 24.

Clock time is called mean time; sun-dial time is apparent time; and the difference between them is equation of time. This will explain the term used in calendars and almanacs, "clock-fast," "clock-slow."

Atmospheric refraction brings in another error in dialling, since it alters the apparent position of the sun, but this effect is very small and need not be considered in a sun-dial, which is, after all, so seldom accurate throughout the year.

The teaching of mathematics in everyday schools, after centuries of surprisingly little change, in the latter half of the nineteenth century became greatly

Dial-head of Centennial Dial.

altered. Ordinary school-books and very commonplace pupils show that in the early part of that century navigation, land-surveying, and dialling were universally taught. Now, few of the young who go down upon the sea in ships know much of the mathematical side of their calling. When every boy wanted to go to sea, and many did go and became ship's captains as soon as they came of age, the study of navigation meant something. The publication of ordnance maps, with many other influences, has set aside the practical and theoretical study of land-surveying. The study of dialling vanished still earlier, with the multiplication of church clocks and watches.

We find Thomas Jefferson amusing himself during an illness in calculating dials, as told in a letter to Mr. Clay.

"POPLAR FOREST, August 23, 1811.

"DEAR SIR: — While here, and much confined to the house by my rheumatism, I have amused myself with calculating the hour lines of an horizontal dial for the latitude of this place, which I find to be 37° 22′ 26″. The calculations are for every five minutes of time, and are always exact to within less than half a second of a degree. As I do not know that anybody here has taken this trouble before, I have supposed a copy would be acceptable to you. It may be good exercise for Master Cyrus to make you a dial by them. He will need nothing but a protractor, or a line of chords and dividers. A dial of size, say of from twelve inches to two feet square, is the cheapest and most accurate measure of time for general use, and would I suppose be more common if every one possessed the proper horary lines for his own latitude. Williamsburg being very nearly in the parallel of the Poplar Forest, the calculations now sent would serve for all the counties in the line between that place and this, for your own place, New London, and Lynchburg in this neighborhood. Slate, as being less affected by the sun, is preferable to wood or metal, and needs but a saw and plane to prepare it, and a knife point to mark the lines and figures. If worth the trouble, you will, of course, use the paper enclosed; if not, some of your neighbors may wish to do it, and the effect to be of some use to you will strengthen the assurances of my great esteem and respect."

Centennial Sun-dial.

Many boys even in that century studied dialling at school — not so many in America, for we had our cheap Connecticut clocks so early in our history, and used them so widely; but in Great Britain dialling was taught, especially in Scotland. Burns says in an autobiographical letter, that he was sent to a noted school "to learn mensuration, surveying, dialling, etc." Hugh Wilson at the same time not only learned dialling, but practised it, and one of his dials still may be seen at Fenwick, near Falkirk. The art of dial-making seems to have been the mode at certain times in various localities. Perhaps some man of note or influence awakened a special interest in the vicinity of his residence. Sometimes a gravestone cutter with a pretty taste for novelties turned his tools to dial-making on dull winter days.

The student was not haled on by old mathematical treatises with any thought of its being an easy task to make a sun-dial. He was warned that he must have "skill in spheriques, together with the laws of Motion of the great Luminaries; he must be absolute in all Circular Affections, as Declinations, Right and Oblique Ascentions, Altitudes, Amplitudes, Azimuths, Culminations, Arches Diurnal, Ascendent, Descendent, etc." All these are not simple of speech nor simple of knowledge to beginners, even if they were entitled, *Easy Rules for Dialling; Young Men's Aid in Gnomonics.* Still, somehow, men of ordinary education, such as stonemasons and men with scant mathematical knowledge, used to be able to make precisely perfect sun-dials from these common rules. I must confess frankly

I find it difficult to understand many of the rules which I have seen, and I have had as good an education in mathematics as have women in general, and as have many men, having studied thoroughly and with great interest algebra and geometry, and having proved myself one of so capable a class in trigonometry that our enthusiastic teacher led his class of girls well on into the knowledge of land-surveying. It does not lessen the value of the education received when I state that this teacher of these girl-mathematicians was a good-looking young man who had been graduated but a year or two from Harvard. In spite of the antiquated diction, I like the rules in Leybourne's *Dialling* better than more modern rules, but Leybourne's book is rare; and even the fine rules for dial-making given in the *Encyclopædia Britannica*, which seem to me perfectly clear and easy of comprehension, have been stigmatized to me within a week as "blind, confused things."

As some will read this book who will care for sun-dials, and perhaps will like to construct one, but cannot understand the application of any rules such as those of Leadbeater or Ferguson, I will give plain rules for making a horizontal dial, worded by H. R. Mitchell, Esq., of Philadelphia, though he disclaims all originality in their construction and phrasing. They seem to me perfect in their simplicity and exactness of information, and with the attached Figures 1, 2, and 3, must be readily understood; and when these are grasped, I am sure the rules for vertical dials, given in the *Encyclopædia Britannica*, will at once seem clear and applicable.

Rules for making a Horizontal Dial

To lay out the lines for a Horizontal Sun-dial in its simplest form, we will first make the Stile or Gnomon.

Figure 1. Lay off the horizontal base line, then with a Protractor ascertain the angle which must be the same as the Latitude of the place, say for New York it is about 40° 44′. Starting from the point *B*, mark the angle with a line *BC*, and draw also the perpendicular line *AC* at right angles to the base *AB*. This gives you the triangular Gnomon in its simplest form; they are generally cut away at the back in a fanciful outline, as indicated by the dotted lines, always leaving the line *BC* untouched, for its line is the shadow line.

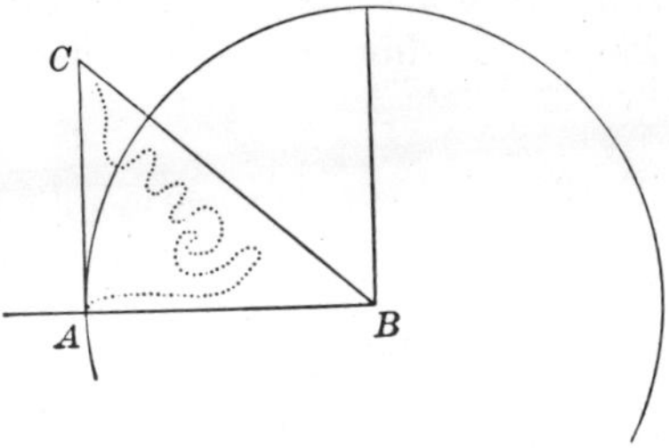

Figure 1.

Then the angle *ABC* will be correct for this Latitude.

For the Face of the Dial draw the line *AB*, and then the line *CD*, at right angles to *AB*. The points *C* and *D* will be your six o'clock points.

At the intersection of these lines, *E*, draw a circle, the radius of which will be equal to the length of the line *BC* in Figure 1.

Then another circle inside, the radius of which will be equal to the length of the line *AB* in Figure 1. Now the outside circle, *A* to *D* and *A* to *C*, divide in two equal parts, and then subdivide each of these parts into three, so as to make twelve equal parts, indicated by *ooooo;* and do the same with the inner circle, indicated by zzzzz.

Now with a rule draw lines parallel to *CD*, from each

of the points of division, *o*, in the two upper quarter circles, then draw lines parallel to *AB* from each point, *z*, on the inner circle.

Marking the points of intersection, then from the central point, *E*, draw lines through these various intersections, and where these lines cross, the circles will be your hour points.

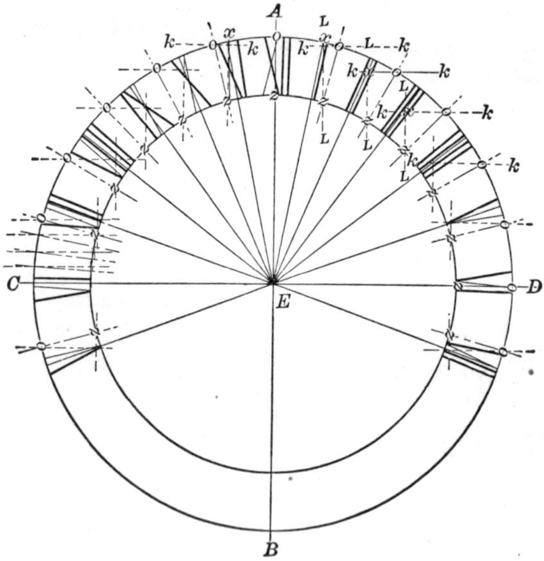

Figure 2.

In drawing the figures for the hours, they should have the same inclination as the lines radiating from *E*. The half and quarter hours should be made in the same way, by dividing the distances between the points on the outer and inner circle, and where the lines from *E* intersect, will give you the marks for the halves and quarters. The minutes, if you choose to put them, can be spaced off with the eye;

the lower circles can be figured in just the same way as the upper, extending the hour marks to, say, 4 A.M. and 8 P.M., but for any ordinary practical use 6 o'clock in the morning and 6 o'clock in the evening covers all that is needed.

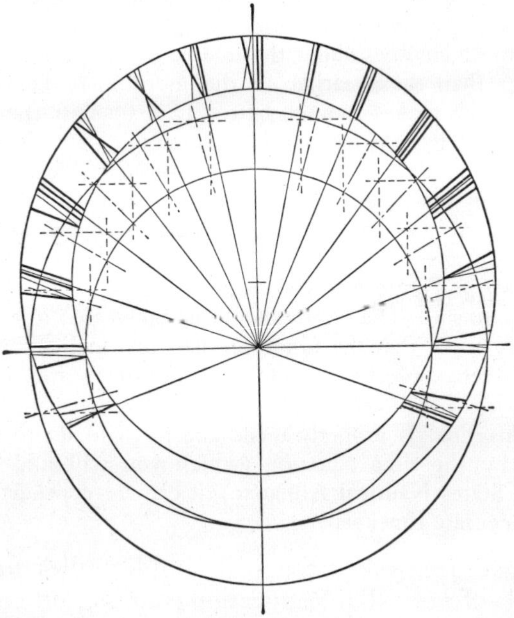

Figure 3.

In laying out a Dial in this way, no allowance has been made for the width or thickness of the Stile or Gnomon; if a thin Gnomon, say, $\frac{1}{16}$ of an inch, is used, it is hardly necessary to make any allowance; but if you want to use a heavy Gnomon, say, $\frac{3}{16}$ or $\frac{1}{4}$ inch thick, then instead of one line *AB*, there must be two parallel lines the same distance apart as the width of the Gnomon, and instead of striking the circle from the central point *E*, two semicircles must be made with the centres where the line *CD* intersects the two parallel lines, or what is probably an easier method would be to cut the draft into two equal parts along the line *AB*, and place between them a strip of paper just the width of the proposed Gnomon.

As the hour lines about the middle of the day are closer together than those earlier in the morning or later in the afternoon, it makes a much better Dial to shift the centre toward the 12 o'clock mark, and draw a new circle from this point, and the lines radiating from *E* can be extended to this new circle, and the Stile or Gnomon will have to be increased in proportion; but the point *must not be moved*, it must always be on the line between the two 6 o'clocks, as shown in Figure 3.

In setting the Dial, care should be taken that it is always perfectly level, that the Gnomon does not incline to either right or left, and should point to the true north (not the magnetic).

If this Dial is properly made and set, and its time corrected by the time equation (which you will find in the United States Nautical Almanac), it can be depended upon as an accurate timekeeper.

I have been asked for still simpler rules for use by folk of very slight education; among others, the negroes of the South,— not the negroes of the better schools, but those who know simply a little arithmetic and a little geography; the geography need be only enough to tell the dial-maker the latitude of his home, which I assume to be the spot where he is to place the dial.

Take a piece of pine board a half-inch thick and fourteen inches square. This will do to make a twelve-inch dial. Paint it with several coats of good oil-paint. Make the gnomon from another piece of board half an inch thick and six inches square.

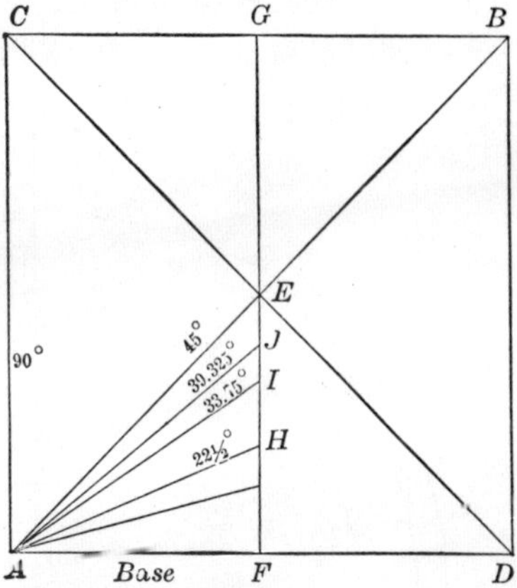

Figure 4.

To shape the gnomon draw lines on Figure 4, from corner to corner, like *AB* and *CD* which make *E* the centre. Then draw the line *GF* passing through *E*, making this line parallel to lines *AC* and *BD*. Of course the angle *CAF* is a right angle of 90 degrees, and the line *AE*, being half of it, makes *EAF* an angle of 45 degrees.

It should be easy to find other angles by dividing these spaces equally. A very easy way is to divide it into divisions of 100 and these subdivide into 50, and so on as shown in Figure 5.

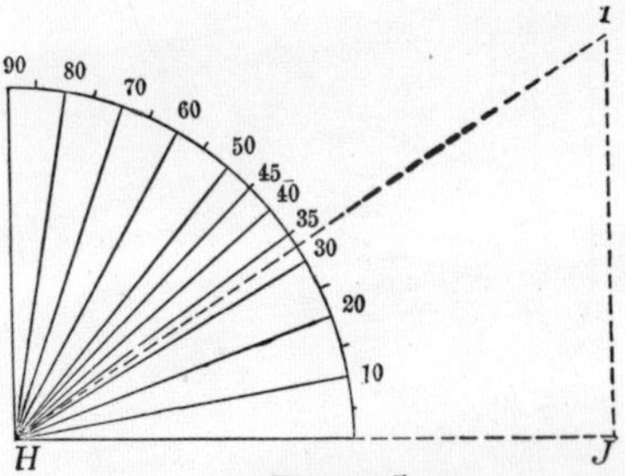

Figure 5.

Let us make a gnomon for Tuskegee, Alabama. The latitude of Tuskegee is about $32\frac{1}{2}$ degrees. The dotted line *HI* is drawn on the angle of $32\frac{1}{2}$ degrees, and the figure *HIJ* forms the proper shape for a gnomon for a sun-dial to be used at Tuskegee.

Now turn to Figure 6.

On the large square of board find as nearly as possible the

centre and draw from it three semicircles marked *E*, *F*, and *G*. The outer circle, *E*, is to have a radius of 6 inches, the next circle, *F*, a radius of $5\frac{3}{4}$ inches, and the circle *G* a radius of 5 inches. Then draw three similar semicircles

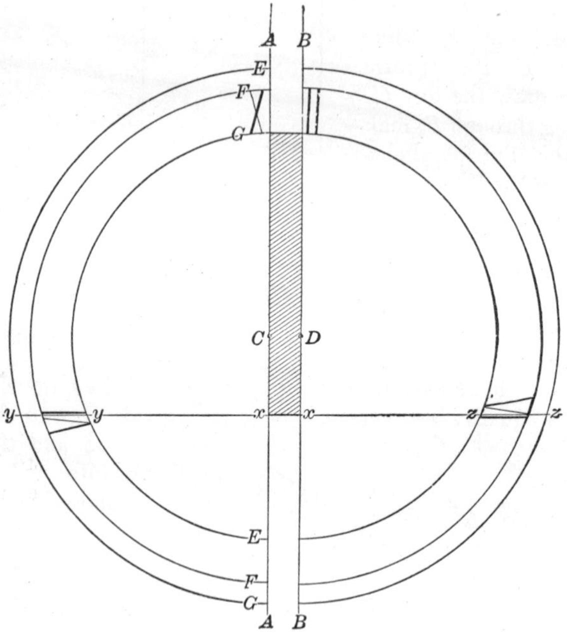

Figure 6.

having *D* for a centre. *C* and *D* must be $1\frac{1}{2}$ inches apart. Between the circles *E* and *F* you can draw in small marks the lines for the hours, half-hours, quarters, and minutes, while between *FG* you will draw the large hour figures which can be plainly seen.

The outside circle *E* will be about $12\frac{1}{2}$ inches in diameter.
The middle circle *F* will be about $11\frac{1}{2}$ inches in diameter.
The inside circle *G* will be about 10 inches in diameter.

We now must set the triangular gnomon *HIJ* of Figure 5 in its proper place. Place it on the dial-face so the point *H* is at *xx*. *J* will then reach the circle *G*, and that point *G* should be numbered XII, the noon hour.

Draw a straight line *yyzz*, across the dial-face running through *xx* (where the point of the gnomon meets the face). This line *yyz* and *zzx* are the 6 o'clock morning and evening hours.

Fasten the gnomon firmly in place by screws from underneath, or in any way you prefer. And then place your dial-face in absolutely horizontal position, and with the gnomon pointing exactly north. Put in the other hour lines by using a good watch keeping correct time, and making the line of the gnomon at precisely 1, 2, 3 o'clock, etc.

Of course these hours can be carefully calculated, but this set of rules is not given for persons capable of such precise calculations. And certainly no excuse is necessary for availing oneself of the easier method obtained through consulting a watch when the *Encyclopædia Britannica*, in its elaborate rules for dial-making, advises the same thing in certain cases. There should also be allowance made for equation of time, if the dial-maker knows how to do so. But without these perfecting details this sun-dial will be found an excellent timekeeper.

While every detail of a sun-dial must be exact, I wish to draw special attention to the importance of the precise shaping and setting of the gnomon;

for great ignorance about the gnomon is displayed even by architects. Upon a house which stands in a neighborhood filled with mathematical interest, near the home of Godfrey, the inventor of the seaman's quadrant, not far from the influence of Franklin, lover of dials, and under the magic spell of the sun-dial of Ahaz, and, moreover, in the only locality in the United States where sun-dials can be found in any number — upon this house is erected a fine dial-face, which even at first glance looks unnatural to you; then you note quickly that the gnomon is a brass or iron bar standing at right angles to the dial-face, upon which are marked correctly lines and numerals for a vertical dial for that latitude. Though this dial with its absurd stick of a gnomon had been erected for some years, I persisted in inquiring until I learned that the dial had been made in London, and on its way hither the gnomon was lost; so the house architect "made a new one," and the man added: "It doesn't matter, anyway; it doesn't keep any time, it's only an ornament. I am told that a sun-dial is never right within sixteen minutes."

CHAPTER V

INGENIOSE DIALLERS

"In this glorious reign, as likewise in the century which has passed, there are to the honor and pleasure of the King and the glory of God in all his works, as seen in the sunne and his motions, many ingeniose diallers."

— *Mathematick Rules* by I. N. Gentn, 1646.

"Methinks it were a happy life
To carve out dials quaintly, point by point."

—*3 Henry VI*, William Shakespeare.

AS CLOCKS and watches became more general they were doubtless often imperfect and harassing in their workmanship. Dutch and German horologers had been warmly welcomed to England, and throughout Europe, but Shakespeare gives a striking example of the carelessness of their work in his taunt: —

"I seek a wife!
A woman, that is like a German clock,
Still a-repairing, ever out of frame,
And never going aright!"

The most celebrated dial and clock maker who came to England was Nicholas Kratzer. In a letter written in English to Cardinal Wolsey from Lucca in 1520, he is called "an Allemagne," a "deviser of the King's horologies," and it was stated that he was "ready to go to England." He was born in Munich, educated in the university of Cologne and other German universities, and became a Fellow of Corpus Christi College, Oxford, in July, 1517. Under the command of Henry VIII he studied astronomy there, and wrote two scientific books which are still preserved in the College Library. Kratzer and Holbein were both men of merry spirit, and both sojourners in a strange land, soon were *arcades ambo;* and the portrait of the dialler here copied (now in the Louvre) was painted by the German artist, and is deemed by many his finest work. In it "Master Nicholas" leans contemplatively over a table with a chamfered scale inscribed: *Imago ad vivum effigem expressa Nicolai Kratzer*, 1528.

Kratzer lived into the reign of Edward VI. At his death many of his papers went to the hands of the astrologer, Dr. Dee, who was an excellent astronomer and dialler. Holbein also was interested in clock and dial making. He designed a time-meter as a gift for the king, which had on its summit a clock driven by wheel-work; below were forenoon and afternoon sun-dials, and still lower a clepsydra.

The most interesting English portable dial in existence is now owned by Lewis Evans, Esq., of Russell Farm, Watford, England. It was once the property of Cardinal Wolsey; and it is without doubt the only existing dial made by Kratzer; in

Nicholas Kratzer, Deviser of Horologies to King Henry VIII of England.

design and workmanship it is worthy that great master. On this page and page 106 are shown a side and a front view of this exquisite dial. The instrument consists of nine dials arranged on a hollow block of gilt brass, $3\frac{5}{16}$ inches high, $2\frac{1}{2}$ inches wide, and $1\frac{9}{16}$ inches thick. The shape of each side plate (as shown in the figures) is (in general) octagonal set upon a base shaped somewhat like the lower half of an equilateral triangle. Seven of the dials are on the side plates or faces which connect these two octagonal plates. There are, therefore, four vertical dials — north, south, east, and west dials — an "upper polar dial," an "inferior polar dial," an "upper equinoctial dial," an "inferior equinoctial dial," and on top a horizontal dial, and a depressed circle which once contained a compass. The tiny gnomons are all formed of thin triangular metal plates.

Side View of Cardinal Wolsey's Dial.

On the lower part of each of the side plates are engraved shields. One (shown in the illustration) bears the arms of Wolsey: sable on a cross engrailed argent, a lion passant purpure between four leopards' heads azure; on a chief or a Lancaster Rose between two choughs. On the other shield is a representation of the arms of the cathedral church of York. Wolsey was made Bishop of York in 1518, and died in 1530, so the dial must have been made in those inclusive years.

Soon after Kratzer was made a Fellow of Corpus Christi he "was made by Cardinal Wolsey his Mathematick Reader." And in the garden of his college this famous "deviser of horologies" put up for King Henry VIII a dial, which has long since vanished. But a careful drawing of it exists in a manuscript by one Hegge, which is now in the College Library; and this drawing is reproduced on page 107 of this book. It bears an extraordinary resemblance to the Wolsey dial owned by Mr. Evans, both in general shape and in position of the dials. And the arms of the church of Winchester are shown in the drawing placed as the arms of the church of York are on the portable dial. Mr. Evans also notes the similarity of the inclination and type of the numerals, the arrangement in narrow circles of the hour numbers, and the resemblance in shape of the gnomons. Both dials resemble a wood-cut in the first book on dialling that was ever printed, a treatise by one Munster; and the portrait of Kratzer by Holbein, facing page 104 of this book,

Full View of Cardinal Wolsey's Dial.

also shows the great dialler with a similar block of dials in his hand. All this certainly proves the assertion that it is the most remarkable portable dial in existence, and cordial thanks are due to Mr. Evans for giving knowledge and sight of it to us all.

Hegge describes Kratzer's dial at Corpus Christi thus:—

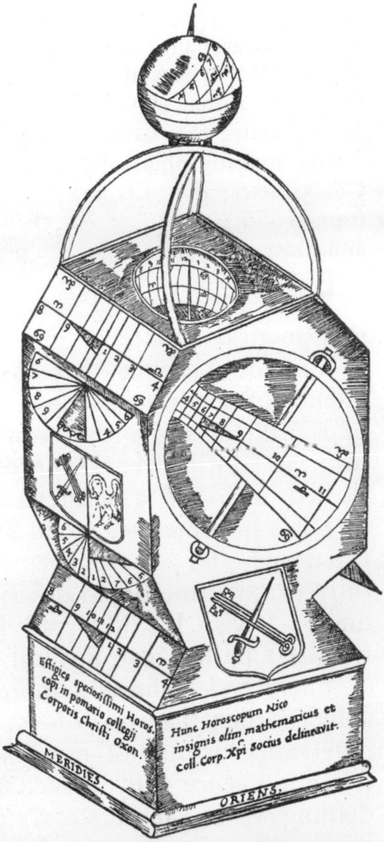

Drawing of Kratzer's Dial, in Manuscript of Robert Hegge.

"In this beautiful Alter (on wch Art has sacrificed such Varietie of Invention to the Deitie of the Sun) are twelve Gnomons, the Sun's fellow travellers, who like farr distant Inhabitants, dwell some under ye Aequinoctiall, some under the Poles, some in more temparat Climats; some upon Plains in Plano; some upon the Mountains in Convexo; and some in the Vallies in Concavo. Here you may see the Aequinoctiall Dial the Mother of ye rest, who hath the horizons of the parallel Sphere for her dubble Province, which suffer by course and half-years night: There the polar Dial wing'd with the Lateral Meridian. Here you may behold the two fac'd Vertical dial which shakes hands with both Poles. There the Convex dial elevated in triumph upon 4 iron Arches. Here, lastly, the Concave Dial which shews the Sun at noone the hemisphere of Night. In other dials neighboring Clocks betray their Errours; but in this Consort of Dials informed with one Soul of Art, they move all with one Motion, and unite with their Stiles the prayse of their artificer."

There is something most touching and stirring in this poetical tribute of one dial-maker to another! Who would think that a treatise on the sombre science of gnomonics could show such an outburst of sentiment and enthusiasm? It is genuine praise, too, the adoring veneration of one craftsman for the skilled hand of a master. The phrase "Consort of Dials informed with one Soul of Art" is an unusual one, and a most felicitous one; it is a beautifully exact term, too, to describe Kratzer's wonderful dials. I wish I could read Robert Hegge's whole treatise; he has won my full respect.

There was infinitely more enthusiasm on such subjects then than nowadays; all sciences were new; diversions for men of mind, for men of parts, were few, and science study served as pleasant occupation; dialling was a science closely allied in the minds of nearly all to astrology as well as astronomy. All had an interest, and nearly every one had a profound belief, in astrology; it influenced many sciences besides that of medicine, and was a favorite study. It occupied with many persons of leisure the place that a study of literature has to-day with the added zest ever clinging to aught mystic; dialling shared in the zest and in the magic.

A belief in the occult influence of the stars and suns and planets upon daily life was universal. Even the few scoffers who dared doubt the validity of a horoscope still heeded the influence of the planets in the humble things of life,—in all farm-work and domestic labor especially. Medicine and astrology were so allied that the soberest dispensatories and medical treatises mingled their rules and influences just as freely as did the old woman who gathered herbs in the full-o'-the-moon. The dial-maker had, therefore, a certain honor cast upon his work because it was allied to still deeper thoughts and beliefs.

In Loggan's *Views of the English Universities* many English sun-dials are shown as they appeared in 1688. Among them is a pillar which stood on the churchyard wall of St. Mary's at Oxford. This pillar bore a cubical stone with dials on four sides, and was crowned by a pyramid surmounted by a ball. This pillar with dials was the work of Nicholas Kratzer; and a full description of it is given in his manuscript *De Horologis*, with the inscriptions which were cut in the stone and written on placards. These were curiously frank, even to the extent of telling that the dial-maker was a heavy drinker. The sun-dial is gone, but there still stands in the quadrangle of Corpus Christi College, Oxford, a similar column dated 1581. It is shown in this book on page 110. It is taller and more elegant, but there is the same cubical block with dials surmounted by a pyramid. This is crowned with a pelican on a globe, the crest of the college. Four coats of arms are carved on the cube: those of the

Sun-dial in Quadrangle of Corpus Christi College, England.

founder of the college, Bishop Fox; the royal arms; the arms of the college; and the arms of Bishop Oldham.

On the column is a perpetual calendar and a motto, *Horas Omnes Complecta*. On the pyramid are in Latin mottoes adapted from the Vulgate: THERE IS LAID UP A CROWN OF RIGHTEOUSNESS. THANKS BE TO GOD; THE GRACE OF GOD WITH ME; I EMBRACE ALL HOURS; I HAVE PLACED GOD AS MY HELPER. This dial was constructed by Charles Turnbull, a member of the college, and his initials are cut on it. A good and full description of it is in Fowler's *History of Corpus Christi College*. It is said that the dial pillar was deemed "inconvenient" when the quadrangle was used as a drilling ground in times of threatened invasion, but happily it escaped being razed. This dial has been frequently copied, in whole and in part. The beautiful cross in the market-place in Carlisle has some of its features. A view from an old print of Carlisle is shown in the chapter on the Rosicrucians.

English horologers vied with the German mathematicians in skilful workmanship. We read that John Poynet, another man of admirable learning, presented Henry VIII with a horologium, which, says Fuller, "I might English dial, clock, or watch, save that it is epitheted Sciotericum." This "observed the shadow of the sun," showed, in addition to the hour of the day, the day of the month, change of moon, ebb and flow of sea, etc. Fuller says severely of such mysteries, "Men never were more curious to divide nor more careless to employ their time than now."

We have some excellent pictures of the dial-makers of the seventeenth century written for us by Fuller and by their contemporary, that equally entertaining old fellow, Aubrey. Dial-makers were much esteemed and much feared, and "were well content to be so." Aubrey says, "In those darke times astrologer, mathematician, and conjurer were accounted the same things."

The Professor of Astronomy at Trinity College, Oxford, who made the fine circular dial on the college library, did not hesitate to secure for himself the name of a conjurer by a system of tricks with a confederate.

William Oughtred, the teacher of many "ingeniose schollars," including Sir Christopher Wren, wrote in 1578 a book entitled *Horologiographia Geometrica*, which Wren translated into Latin when he was but fourteen years old. Let me quote from Aubrey's lively account of him: —

"Oughtred was a little man; had black haire and black eies with a great deal of spirit. His witt was always working. He would draw lines and diagrams in the dust. His son Ben told he did use to lye a-bed till eleaven or twelve o'clock, with his doublet on. Studyed late at night; went not to bed until 11 o'clock; had his tinder box by him; and on the top of his bed staffe he had his inke-horne fixt. He slept but little. Sometimes he went not to bed at all in two or three nights, and would not to come downe to meales till he had found the *quæsitum*. Severall great mathematicians came over into England on purpose to be acquainted with him. His country neighbours knew there must be some extraordinary thing within him he was so visited by foreigners. He did not like any save those who tugged and took paines to worke out questions. He taught all free. He could not endure to see a scholar write an ill hand; he taught them all presently to mend their hands.

"He wrote a very elegant hand, and drew his schemes most neatly as they had been cut in copper. He was an astrologer, and very lucky in giving his judgments in nativities: he would say, that he did not understand the reason why it should be so; but so it would happen: he did believe some genius or spirit did help. The country people did believe that he could conjure, and 'tis like enough

Triangular Lodge with Sun-dials, Rushton, Northamptonshire.

that he might be well enough contented to have them think so.

"He was a great lover of Chymistry and told Jno. Evelyn not above a yeare before he dyed, if he were five younger he doubted not to find the philosopher's stone. The olde gentleman was a great lover of Heraldry. His wife was a penurious woman, and would not allow him to burne candle after supper, by which means many a good notion is lost, many a probleme unsolved. Mr. Hanshaw, his scholar, when he was there, bought candle, which was a great comfort to the old man. His son Ben was confident he understood magique."

The old dial-maker died of joy for the coming in of the king. Son Ben is a distinct character, and takes his departure from history, "putting on his considering cap, which was never like his father's, with much adoe to find the place where lie his father's bones; for truly his grief was so great that he could not remember the place." The chronicler says dryly, "Now I should have thought it would have made him remember it the better."

In reading the biographies of men of the sixteenth and seventeenth centuries, we note that their skill in dialling is ever made the subject of much praise. Thus Aubrey writes of one Edward Halley: —

"As a boy he studied Arithmetique and was very perfect in the Coelestial Globe. He studyed Geometry, and at 16 could make a Dyall and then he said he thought himself a brave fellow. When at the age of 19 he solved this useful probleme, never done before, viz.: From 3 distances given from the Sun and Angles between to find the Orbe, for which his name will ever be famous. At 1678 he added a Spectacle-glasse to the Shadow-vane of the lesser Arche of the sea-Quadrant (or back-staffe) which is of great use, for that spot of light will be manifest when you cannot see any Shadow. He presented his Planisphere with a short description to her Majesty who was very well pleased with it; but *got nothing but prayse.*"

Aubrey refers to the wonderful dials of Wren, and also relates of Robert Hooke, the inventor of pendulum watches, about the year 1680, that when but a boy, he made a dial on a round trencher without any mathematical instruction, — which was not remarkable, after all, for a man who invented thirty different ways of flying.

A great number of mathematical works of about this date exist in various public and private libraries; these are often in manuscript, for the market was overstocked. Fuller remarks in his *Worthies of England*, "I never did spring such a covye of mathematicians all at once, as I met with at this time."

The interest in sun-dials in England must have been vastly spurred on by the never flagging zest of King Charles I for them in every form. It was one of the touching stories which I read in my childhood that he ever carried a silver pocket-dial which he gave on the night preceding his execution to his attendant Herbert as a last gift to his son, the Duke of York. I wonder what was the title of that book which I so loved, which gave anecdotes of the English kings, princes, and princesses! There were several of King Charles in it, and they helped to make him the idol of my childhood, — a regard I cannot even now divest myself of, though years of mature reading have forced upon me other tales than the sad and romantic ones of that little picture-book. I had not the slightest notion what a dial was; but since it was associated with the king's dying, I had a wild fancy that it was something, a silver box that contained poison — to steal the old motto-jest — a die-all.

One of the most superb dials ever erected in England was the one in the king's garden at Whitehall, set up for King Charles II. I have a description of this magnificent and singular dial in Leybourne's *Tractates*, but a minute account is in a book written by the maker of the dial, Rev. Father Francis Hall, of the Socicty of Jesus. A copy of this latter rare volume was lent to me by Mr. Lewis Evans. A briefer account may be seen in *The New Universal Magazine* of January, 1756. The title of Father Hall's book may be given in full, as being most explanatory: *An explication of the Diall Sett Up in the King's Garden at London in 1669. In Which Very Many Sorts of Dyalls are Conteined; by which besides the Houres of All Kinds diversely expressed, many things also belonging to Geography, Astrology, and Astronomy are by the Sunnes Shadow made visible to the eye. Amongst Which, Very Many Dialls, Especially the Most Curious, are New Inventions, Hitherto Divulged to None. All these Particulars are Shortly yett Clearly sett forth for the common good. By the Reverend Father Francis Hall (otherwise Line) of the Society of Jesus, Professor of Mathematicks.* The book has sixty-nine interesting illustrations. The *Elevation* of this dial from Leybourne here copied is equal to a description, but some of the curious details may be pointed out. The dial stood on a "Pillar" or "Piedestall" of stone, being in six parts, set in the general form of a "Pyramis." The lower table was about forty inches in diameter, and had twenty dials set around the edge, all covered with glass. Some of these showed the hour after the Jewish, some the Babylonian, some of the Italian fashion. The gnomon of each was a lion's paw or a unicorn's horn. On the upper part of this table were eight reclining dials; these were curious. One showed the time by the shade of the style falling on the hour lines, the next by the shade of the hour line falling on the style, the third had no shadow. Of the four faceted globes attached to this table, one had several dials "belonging to Geography," the second several dials "belonging to Astronomy," another to Astrology. There were also four globes with dials attached by iron arms.

The second table was thirty inches in diameter. This had sixteen dials on the circumference; these differed from those first named, in that the former were drawn on the back of the pieces of glass that protected them, while these sixteen were on the stone. These showed "the different manner of Rising of the Stars to Witt; the Cosmicall, the Cronycall, and the Heliacall." The styles of these dials were little stars painted on the inside of the glass cover. This table had eight reclining dials, four of which were of mirrors which reflected the

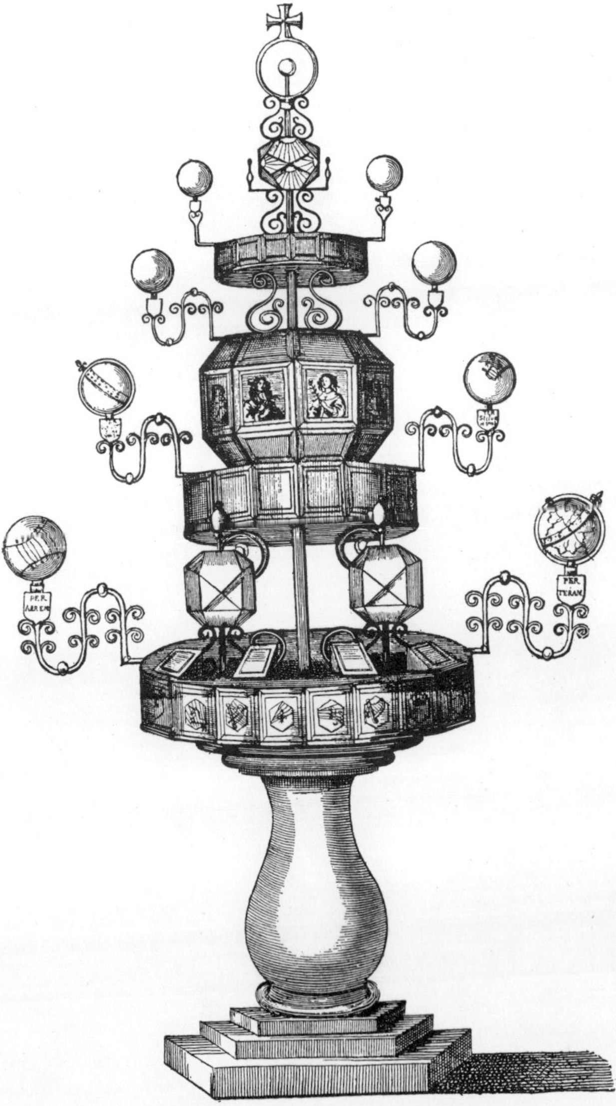

Elevation of Dial of King Charles II at Whitehall.

shadows on dials placed above on the third table. There were four arms with globes attached.

The third piece of the dial was a sort of globe cut in twenty-six faces. Some of these faces were covered with glass which served as windows, letting the observer look at dials within the globe. From this four iron branches held each a glass globe painted within in such a manner that they also served as dials.

The fourth part or table, twenty inches in diameter, was cut on the edge into twelve semicylindrical concave dials, of which the style was a Flower de Luce. The four glass globes on this all were ingeniously varied.

The fifth part, a globe twelve inches in diameter, was cut in faces of which fourteen bore each a dial.

The sixth part was a glass globe seven inches in diameter, held by bands of iron and surmounted by a cross.

This very bald outline conveys no idea of the ingenuity displayed in these many dials, which numbered nearly three hundred, and were of seventy-three different kinds; the four globes attached by iron arms to the lower table were marked each, *Per Ignem*, *Per Aquam*, *Per Aeram*, *Per Terram*. The first bowl, filled with water, showed the hour by the heat of the converged rays. The second bowl, also filled with water, threw a little circle of light on the true hour. The third globe, filled with air, exhibited "two little pictures of the sun like two stars," and thereby marked the hour. The fourth globe had "a geographicall description of the whole earth upon the inward superficies of the Bowle."

The four great globes standing upright on the lower table were very singular; each bore thirty-two dials, of which twelve were pentagons; and they served to show "supper time," "sleeping time," etc., of scores of points all over the world. The globes attached by iron branches to the second table bore

Hexagonal Dials on King Charles Dial.

the titles, PER STYLUM SINE UMBRA, PER UMBRAM SINE STYLO, PER STYLUM ET UMBRA, and SINE STYLO VEL UMBRA, which indicated their character. Through the little peep-holes of the largest globes could be seen pictures, including portraits of the king and queen, queen mother, Duke of York, Prince Rupert, etc.

This old book with its curious wording and illustrations has raised a world of fancies and dreams in my mind. I can see the gay and thoughtless courtiers, and the equally volatile and careless king, bending eagerly over this dial, while Charles pointed out to Nell Gwynne, perhaps, the various singular contrivances which formed the dial-faces, and translated to her the scores of Latin mottoes; for each dial of the hundreds, big and little, had some motto or inscription. It seems a pendant to the pages of Pepys, — a presentment of the court life the diarist so truly loved.

And how quickly it all vanished! like the graphic scene of the king's gay court, and his sudden death, so strikingly told by John Evelyn. So quickly had this wonderful dial disappeared that Leybourne could write, "Thus have I given a brief account of this now demolished Dial, which account and figure together may give some light to the ingenious Practitioner to invent infinite Varieties of this Kind."

CHAPTER VI

PORTABLE DIALS

> "And then he drew a dial from his poke
> And looking on it with lack-lustre eyes
> Says very wisely, 'It is ten o'clock.'"
>
> —*As You Like It*, Act II, Sc. vii, William Shakespeare.

BY SHAKESPEARE'S day many pocket-dials were in use in Europe; the ring-dial and shepherd's-dial were common; the compass-dial less so. The name poke-dial, given to them in old poems and plays, brings a pleasant study of the words *poke*, *pouch*, *pocket*, *purse*. The Latin words *portarium* and *solarium* are also employed for these pocket-dials. Other references are made in the pages of Shakespeare to pocket-dials, among them the shepherd's-dial, as in *3 Henry VI*.

> "Oh God, methinks it were a happy life
> To be no better than a homely swain,
> To sit upon a hill as I do now,
> To carve out dials quaintly point by point;
> Thereby to see the minutes how they run,

How many make the hour full complete,
How many hours bring about the day,
How many days will finish up the year,
How many years a mortal man may live."

Let me tell the history of this poke-dial, and give rules for making it, so that, as do still the shepherds in the Pyrenees, any one may "carve it out quaintly point by point." A shepherd at Béarn made this philosophical answer last summer to an American who jested with him over the antiquated timepiece he was carving: "No human being can certainly dare to know the time of the day as well as the sun, since without him there would be no time; so we go directly to the sun when we wish to know what time it is."

This dial is known by many names old and new: chilindre, cylinder, calendar, kalendar, column-dial, pillar-dial, shepherd's-dial, Pyrenean dial. Treatises on these chilindres are extant which were written in the thirteenth century.

Warton gives as a note to Lydgate: "Kalendar, Chilindre, cylinder, a kind of pocket sun-dial." Chaucer says in his *Shipman's Tale*, "By my chilindre it is prime of daye," which was the end of the first hour after sunrise.

These cylinders are small columns of ivory or wood having at the top a kind of stopper or lid with a ring at the top, and with a gnomon hinged upon the side of the stopper. The cylinder is divided into month spaces on the circumference. When in use, the stopper was taken out and the gnomon turned around, so it hung over the desired month space or line. Then the dial was hung up on the ring, so it hung exactly vertical with the pointer extended toward the sun. It could be set on a horizontal plane, but a slight deviation so affected it that it was far better to hang it up. The

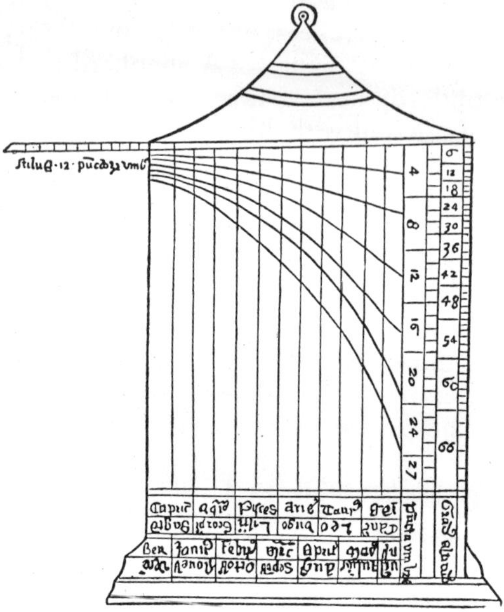

Drawing in 14th Century Mss. of a Chilindre.

shadow fell on the curved hour lines and showed the time. Homan in his *Vulgaria* gave in 1520 a very terse description of these dials, calling them "instruments like a hanging pillar with a tunge hanging out to know ye tyme of day."

On this page is given a drawing from a fourteenth century treatise on the *Chilindre* which is now preserved in the Arundel Mss. It is called in it the "travellers'-dial"; full and precise directions are given, as for the wood to be used, which should be "very solid, imporous, equal, and without knots."

Two Boxwood and One Ivory Shepherd's-dials.

The markings and lines are carefully shown with exact directions for making them. The gnomon is called in this treatise a style or indicator, and could be made of copper, of silver with a bit of lead melted on it; it worked on a pin fastened in the lid.

In the museum at Naples is the oldest portable dial known, and it is an adaptation of the chilindre. It was found at the excavations made in Herculaneum in 1759. It is of bronze shaped like a ham,

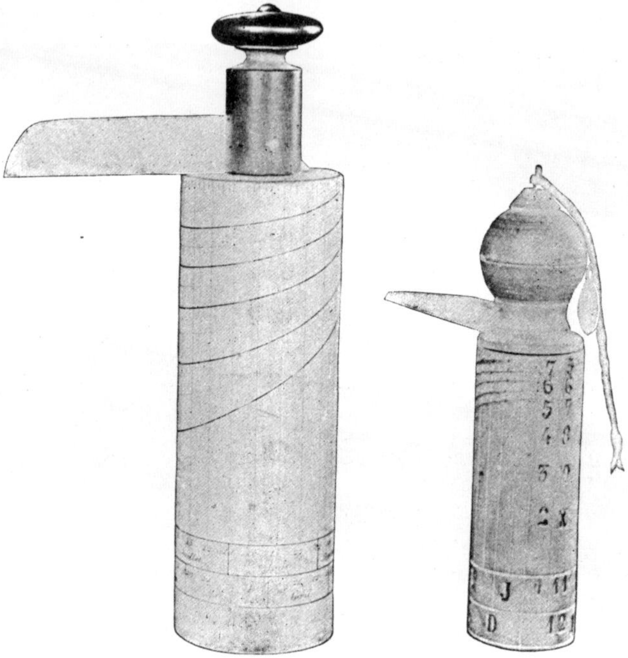

Two Boxwood Pillar-dials.

and on the flat sides are the lines and letters that prove it to be a sun-dial. Its date must be before A.D. 79. It was to be suspended by a ring and had a tail-piece which must have been the gnomon.

A most interesting type of cylinder-dials is known in India, where the dials are set in staves $4\frac{1}{2}$ to $5\frac{1}{2}$ feet long, which pilgrims carry with them to Benares. These are called *Ashadah*, the name of the month when these pilgrimages are usually made, — it is the latter half of June and first half of July. These pilgrim staves are eight-sided, carved with numerals to show the half hours from sunrise to sunset.

On page 123 is given an interesting group of chilindres, owned by Mr. Evans. Number 1 is a shepherd's-dial, of boxwood, $3\frac{3}{4}$ inches high, $\frac{7}{8}$ of an inch in diameter. The figures and month initials were stamped on it, and then filled in with some red pigment. This was bought in 1899 at Argeles, near Lourdes, where these dials are still sold for use. Number 2 is a wooden column-dial 5 inches high, covered with printed paper varnished. Made for latitude 49° by *Henry Robert, horologer au Palais Royal, N° 164, Paris.* Its probable date is 1800. Number 3 is a column-dial of ivory $4\frac{3}{4}$ inches high; marked *J. Le Tellier ÆDieppe.* Date about 1780. On page 124 are shown a column-dial $4\frac{1}{2}$ inches high, $\frac{7}{8}$ inch diameter, probably German; made about 1650; also a shorter boxwood pillar-dial, which once belonged to Mr. Lewis's great-grandfather, Lewis Evans, F.R.S., and which may have been made by him. Its date is about 1780; it is $3\frac{1}{2}$ inches high, $\frac{3}{4}$ of an inch in diameter.

I give here on pages 126 and 127 two plates on the making of a cylinder-dial from Ferguson's *Mechanical Lectures on Dialling.* These plates show how to construct a cylinder-dial for the latitude of London.

Plate I, on this page, was 8 inches long and $7\frac{1}{2}$ inches deep, and was to be pasted around a cylinder 6.60 inches long below the movable top, and 2.24 inches in diameter. The cylinder should be hollow to hold the style when not in use; the style when fixed must be at an exact right angle to the cylinder, and be placed at top of the line *AB* of the Plate I,

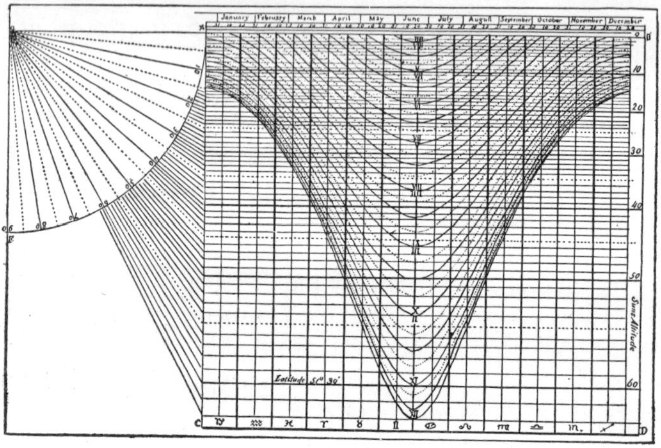

I. Plate for Cylinder-dials.

where the parallels of the sun's altitude begin. The length of the style (or distance of point *C* from the cylinder) must be equal to the radius *aA* of the quadrant in Plate I. The rules for the construction run thus:—

Draw the right line *aAB* parallel to the top of the paper, and with any convenient opening of the compasses set one foot in the end of the line at *a* as a centre, and with the other foot describe the arc *AE*. Divide it into ninety degrees. Draw the right line *AC* at right angles to *aAB*, touching the quadrant at point *A*. Then from the centre *a* draw right lines through as many degrees of the quadrant as are equal to the sun's altitude at noon on

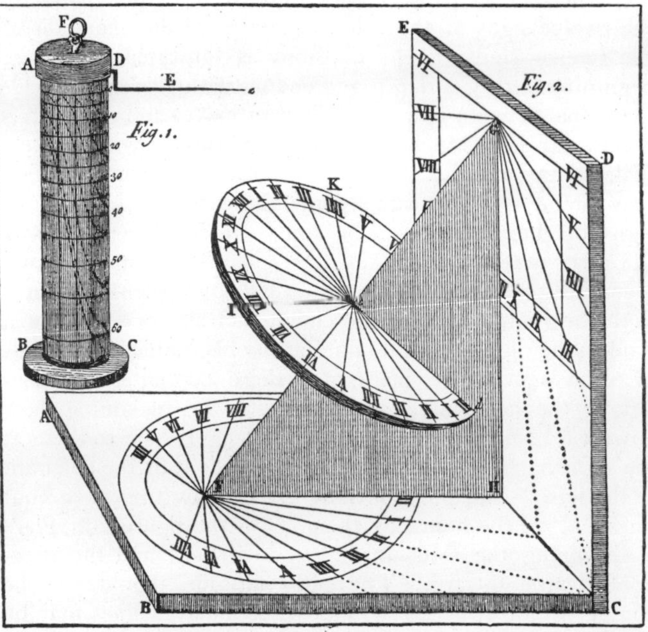

Diagram of Construction of Cylinder-dial.

the longest day of the year at the place for which the dial is to serve. (At London it is sixty-two degrees.) Continue these right lines till they meet the tangent line *AC*. Then from each point of meeting draw a straight line (at right angles to *AC*) across the paper (sixty-two in this place) all being parallel to line *AB*. These lines will be parallels to the sun's altitude, in whole degrees, from sunrise to sunset, on all the days of the year. These sixty-two lines or parallels must be drawn to the line *BD*, which must be parallel to *AC*, and must be as far from it as the intended circumference of the cylinder. Divide the lines *AC* and *BD* by parallel lines into twelve equal parts for the twelve signs of the ecliptic; and place the character of the twelve signs in the divisions as indicated, which is, beginning with Capricorn and ending with Pisces. Divide these spaces again by parallel lines in halves and in quarters if they can be distinct. At the top of this drawing make a scale of the months and days, and place it so that the days may stand over the sun's plane for each of them in the signs of the Ecliptic. The sun's place for every day of the year may be found in any Ephemeris (or almanac). Compute the sun's altitude for every hour when the sun is in the beginning, middle, and end of each sign of the ecliptic. And in the upright parallel lines at the beginning and middle of each sign make marks for these computed altitudes among the horizontal parallels of altitude, reckoning them downward according to the numeral figures set to them at the right hand answering to the like divisions of the quadrant at the left. And through these marks draw the curve hour lines and set the hours to them, as in the figures on Plate I, reckoning the forenoon hours downward and the afternoon hours upward. The sun's altitude should also be computed for the half hours. The quarter lines may be drawn by the eye. Cut off the paper at line *AC*, also at line *BD*, also cut close to the top and bottom horizontal lines. It is then fit to paste on the cylinder.

I have given these rules in full not only that any curious reader may amuse himself by making a shepherd's-dial, but to show the wording of the clearest of these old-time mathematical treatises. The drawing

Standard of Dials, with Compass. From Ferguson's *Lectures*.

of the lines for a cylinder-dial for the latitude of London were engraved on a sheet of strong paper and sold in large numbers.

The story of James Ferguson, the "self-taught astronomer," the author of these rules, and his connection with dial-making is interesting. He was a thoughtful boy, ever occupied in watching the stars

Brass Octahedral Block of Dials.

while he served as a shepherd, and drawing what he called star-papers. His intelligence and application gained the interest of a gentleman's butler named Cantley, who, when the lad was sixteen years old, taught him to make sun-dials. Cantley must have been a remarkable butler, for he was a first-class mathematician, a master of every musical instrument save the harp; knew Latin, French, and Greek; "let blood extremely well, and could even prescribe as a physician upon urgent occasion." We wonder a bit how the family fared whom he served. He was painting a dial on the schoolhouse wall when Ferguson saw him, and soon taught the eager farm-laddie algebra, astronomy, and dialling. With working in a mill, painting portraits, drawing designs for embroidery, making clocks and dials, Ferguson soon was as busy as Cantley. The years passed, and suddenly Ferguson had made his name as a writer, lecturer, and inventor in gnomonics. He invented in 1766 a "Universal Dialling Cylinder," by the use of which all kinds of sun-dials could be easily calculated and made. These he made and sold, as he did orreries, cometariums, astronomical clocks, and sun-dials. He had a mechanical rather than a mathematical mind; and it may uphold us in our mechanical rather than mathematical modes of constructing sun-dials to know that Ferguson never was able to understand Euclid, and his constant method of satisfying himself of the truth of any problem was by measurement with a scale and pair of compasses, not by mathematical demonstration. Ferguson made many

Wooden Block Dial, with Paper Figures; 1780.

dials of various kinds to use in his lectures on dialling. An interesting one of his inventions is shown on a succeeding page.

The astrolabe was well known in Persia, Arabia, and India at an early day. Originally the word was

Brass Equinoctial Dial.

applied rather vaguely to any flat circular instrument used for observation of the stars; then it was restricted to the kind called the astrolabe planisphere, — I will not describe an astrolabe, as reference to any scientific dictionary or cyclopædia will afford an exact description. With various contrivances to use as a gnomon, they served as portable dials for many centuries; such was the astrolabe of Chaucer.

I suppose no one to-day ever reads Chaucer's *Treatise on the Astrolabe*, or *Bread and Milk for Babes*, save the painful editor, Rev. Mr. Skeat. But a superficial glance at his rendering of it shows a very gentle and pleasing trait of our beloved poet; a side of his character not unexpected to those who love his works, but in some way unexpected in its power of moving your sentiment. We are bidden to believe that the English of that and the two succeeding centuries showed much severity to their children; but the absolute memorials which we have of intercourse between parents and children do not, to me, prove the assertion.

Chaucer was pleased and proud of the progress of his little son in mathematics, which, with "the Languages," formed the substance of schooling in that day. He had previously given the child a small astrolabe as a reward, and the child wished to learn about it; and as Latin treatises were hard to comprehend, the father wrote one suited to the child's mind. Here are his own words: —

"Litell Lowys my son, I have perceived well by certain evidences thine abilitie to lerne sciencez touching nombres and proporcions; and as well, I consider thy bisi preyer in special to lern the tretis of the Astrelabie. Than, for as much as a philosofre seith, he wrappeth him in his friend that condescendeth to the rightful preirs of his friend, therefor have I given thee a sufficient astralabie for owr orizonte [horizon] compared after the latitude of Oxenford upon which by inditecon of this litel tretis, I propose to teach a

Armillary Sphere-dial, Brockenhurst, New Forest, England.

certain nombre of conclusions apporteying to this instrument."

His conclusions are interesting: that the astrolabe was the most noble of instruments; that they were too little known; and that all treatises upon them were "too hard to thy tendre age of X years to conceive."

"I wil showe litle reules & naked words in Englissh since Latyn he kanstow yet but smal, my litle son.

"And I pray trewly every discreet person that redith this litle tretis to have my rude endyting excused, and my su-

Brass Block Dial made in Styria, Austria, 1691.

perfluietes of words for two causes. In the first for that curious enditing & hard sentence is ful hevy atones for swich a child to lern. & the second cause is this; that sothly me seemeth better to write unto a child twice a good sentence than he for-get it once."

After all this declaration of his determination to write simple rules, it is a disappointment to read the rules themselves. I cannot see how any child of ten could have possibly understood them, even a fifteenth century child. The description of the astrolabe is, however, very clear.

There is shown and described on pages 105 and 106 the most interesting English portable dial known. It was made by Nicholas Kratzer for Cardinal Wolsey, and is now owned by Lewis Evans, Esq.

In the latest edition of Mrs. Gatty's (now Mrs. Eden's) *Book of Sun-dials* is a valued addition, a chapter upon *Portable Sun-dials*, written by Mr. Evans. Certainly no one else could be so well fitted to write on this subject as he, since he possesses the finest collection of portable dials in the world. His work on Mrs. Gatty's book has been equalled by his kindness to me in my writing of portable dials in this book, both in furnishing me with illustrations and in giving me the history of special examples. We have in America at the National Museum at Washington, D.C., a small but good collection, of which the best are displayed in this book, through the kindness of various members of the staff at the Smithsonian Institution, who had special photographs taken for me. Mr. Evans owns over 450 portable sun-dials besides astrolabes and like instruments. He turned to collecting dials, since they seemed to be the only thing that was not being collected in his family. Among the objects are British, Greek, and Roman coins, flint and bronze implements, etchings, china, posey-rings, stamps, bank notes, shells

— land, fresh water, and sea shells, butterflies, moths, birds' eggs, fossils, etc. His elder brother, Arthur John Evans, F.R.S., is the head of the famous Ashmolean Museum at Oxford, and collects all such objects, and in addition Greek and other engraved gems. The taste for collecting rarely runs in a family; few sons care for the collections of their fathers, so the extent and vehemence of this family trait is surprising. Mr. Evans has the finest library on the subject of dialling that has ever been gotten together. Over four hundred books and pamphlets upon dialling; these in many languages. His interleaved copy of Mrs. Gatty's *Book of Sun-dials* must be a wonderfully interesting thing. There is another collection of dials in England, owned by a Mr. Fry of Bristol. Mr. Kent, the artist, has a collection of garden-dials at Hayward's Heath.

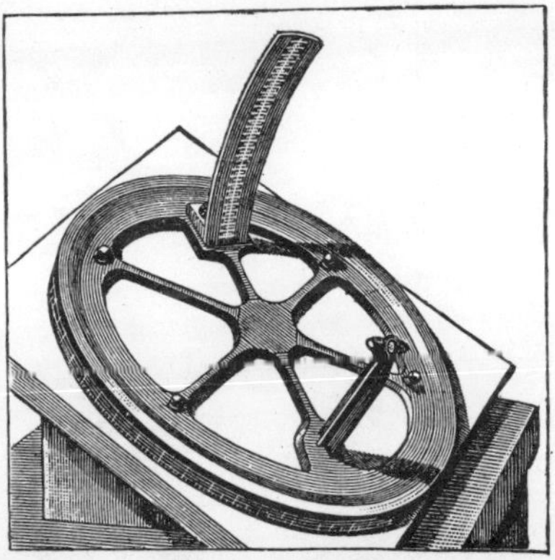

Thevenot Sun-dial.

The British Museum has the best public collection of portable dials; the Victoria and Albert Museum at Kensington has also a few dials. The best collection in Germany is at the Physical Museum at Dresden, and the German Museum at Nuremberg; in Italy, the Galileo Room of the Institute of the

Studii Superiori in Florence, and the College Romano at Rome. There are a few dials in the Musée Cluny, also one fine private collection in Paris. I know no

Brass Universal Ring-dial, set for Use; owned by Author.

private collection of portable dials in America save the few I own.

Readers of this book owe to Mr. Evans a debt of cordial thanks; for through his generosity I am enabled to make this chapter the most fully illustrated chapter on portable dials ever printed in English, or I believe in any language; and illustrated, too, with the rarest and most beautiful examples of their kind. Many are unique; and all would be beyond any possibility of examination and comparison, save for his thoughtfulness and kindness.

Brass Universal Ring-dial, Flat for carrying; owned by Author.

An ancient form of portable dial was called the ring-dial; it is old as Shakespeare's day. These varied greatly in size. Mr. Evans has seen but one English ring-dial small enough for a finger ring; though he has them of the size proper to use on a watch fob or as a seal. These were from $1\frac{1}{2}$ inches to $2\frac{1}{2}$ inches in diameter and were made in large numbers in Sheffield throughout the eighteenth century. In this form they were simply a flat ring with the hour lines drawn diagonally across the narrow surface of the ring. Through a tiny hole drilled in one side the sun's rays sent a shining spot of light upon the hour lines.

On page 136 is shown, full size, a "universal ring-dial," owned by the author. This is a very good example of a common form of this dial, which was greatly used in England in the seventeenth and eighteenth centuries. It is, as may be seen in the illustration, a kind of armillary sphere showing the circles of the equator and the meridian together with the polar axis (the bar across the centre), which is marked with the names of the months and has on it a sliding gnomon with a tiny hole in the centre. The whole is pivoted together in such a manner that it can be flattened out (and is thus shown on page 137) and can be carried conveniently in the pocket. The meridian ring is marked with degrees, and the suspending clip is movable so it can be placed in proper position according to the latitude. When in use, it is held up by the ring and turned until a little line of light falling through the hole in the gnomon strikes on the centre line of the hour circle, and thus tells the correct time. This hour circle is marked with two sets of numerals, one for forenoon and one for afternoon. The gnomon must of course be carefully set to the

Universal Ring-dial, with Disc Axis.

position proper for the month of the year, as shown on the table of months marked on the axis. Another form of ring-dial is on the opposite page.

Three full-sized figures of interesting ring-dials owned by Mr. Evans is given on this page. One

Three Brass Ring-dials.

is a brass ring-dial $2\frac{1}{8}$ inches in diameter, made about 1760 by Proctor, and inscribed: —

Set me	Use me	And I
Right	Well	Time tell

Number 2 is a smaller brass ring $1\frac{3}{8}$ inches in diameter, called a seal-dial. This is an English dial dating about 1560. The seal has the initials A. P. and an oak branch. Number 3 is a German dial, date, 1698, differing from the ring-dial carefully described, in that it has no sliding ring, but has two holes opposite each other, one for use in winter, one in summer. On this page is shown an English signet ring of silver, of about 1620. The crest is a dog. The dial is less than an inch in diameter. Its companion is a brass ring-dial, either Swiss or German; $1\frac{9}{16}$ inches in diameter, and the ring is $\frac{3}{8}$ of an inch wide. This differs also from the common type in having only one set of hour lines, stamped on an extra piece of brass, curved, within the ring.

Brass Ring-dial and Silver Dial-ring.

Posies suitable for rings were deemed also well fitted as mottoes for ring-dials. Mr. Evans has in his collection dials with these familiar old lines:—

"A Ring is Round and Has No End
So is My Love Unto My Friend."

"The Love is True: That I O V
As True to Me: Then C V B."

"As Time and Hours Pass Away
So Doth The Life of Man Decay."

In Exeter Museum is a ring-dial with the motto which I have seen on an old stoneware Georgius Rex jug:—

"If My Master Use Me Well
I'll Try All Others To Excell."

"A Stitch in Time
Saves Nine"

would seem incongruous for a dial, had I not also seen it on a tea-pot.

Mottoes are, however, comparatively rare on ring-dials, for there was scant space for such engraving, especially when a list of large cities with their several latitudes was given.

Universal Ring-dial, with Base and Feet with Screws.

This form of ring-dial has been used in an enlarged size, but of same general shape, fixed on a block to stand on a window-sill or table; one with base and adjustable screws is shown on this page. Modern ones, such as the one on page 142, are now sold in London at the shop of F. Barker & Son, Clerkenwell Road. Larger still, in one case with rings a yard in diameter, this form of dial has been seen as a garden-dial. One such, symbolizing the earth, is borne on the head of an Atlas of stone at Oakley Park in Shropshire. A very beautiful adaptation of this form is seen in the garden-dial at

Brockenhurst, New Forest, shown on a later page. Another is at Clumber, the home of the Duke of Newcastle, and still another at Holland House. The shadow of the nodule on the axis falls on the inside of the circular band, which is elaborately lettered and numbered, often in gold. This great armillary sphere, seen against the green background of trees in a beautiful garden, is a most noble sight. I have always longed to have in my library a splendid great revolving globe, such as are seen in our higher schools; equally do I covet one of these sun-dials patterned upon the armillary sphere. An equally delightful mathematical instrument, one which makes us know the beauty of numbers, of pure mathematics, is a Sunshine Recorder; a sun-dial which records every second of sunshine. Their beautiful great clear globes are a joy to behold, and have all the mysterious charm of a Chinese crystal globe. They are costly, for they are perfect things of exquisite workmanship. One from the firm of

Globe Window-dial. F. Barker & Son, London.

F. Barker & Son has just been set up in Washington.

It has been frequently asserted that George Washington was in the habit of carrying a pocket-dial instead of a watch, and I have seen a silver pocket-dial which was given him by Lafayette. He certainly had watches in plenty, as we know by the many well-authenticated ones still existing which once were his. I also know of a pocket-dial that was

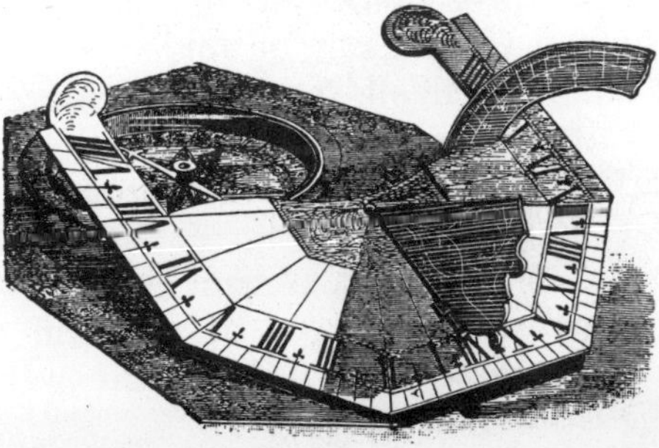

French Pocket-dial, with Compass. Le Maire.

carried by Rochambeau. They were evidently the mode among Frenchmen of rank and station of that day. There is also in the collection of the Washington Association of New Jersey, in their spacious house at Morristown, New Jersey, a very dainty, elegant pocket-dial of silver, the card label of which is marked: —

"Silver pocket Sun-dial used by a French officer in the Revolutionary War, made in Paris in 1673. Presented by Mr. Henry W. Miller."

On page 143 is shown a French pocket-dial of different shape, — a silver pocket-dial with compass, made by Le Maire in Paris two hundred years ago. When a plummet is added, this is deemed by modern dial-makers the best of all the pocket-dials, for use by soldiers, colonists, and travellers.

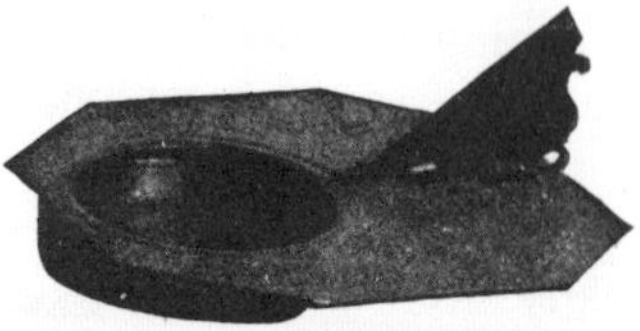

Silver Portable Dial, French.

It is so hinged that it can be folded very flat and thin, and takes but small space in the pocket. The names of various towns with their latitudes are given, and the arm is so marked that the proper angle for the dial-face is easily found.

The choicest silver dials made in France were often furnished with shagreen or fish-skin cases to protect them, as were the watches of those days.

These fine French pocket-dials came in various shapes, — oval, round, octagonal, oblong; and in Italy, a cross-shaped dial was made, holding a reliquary.

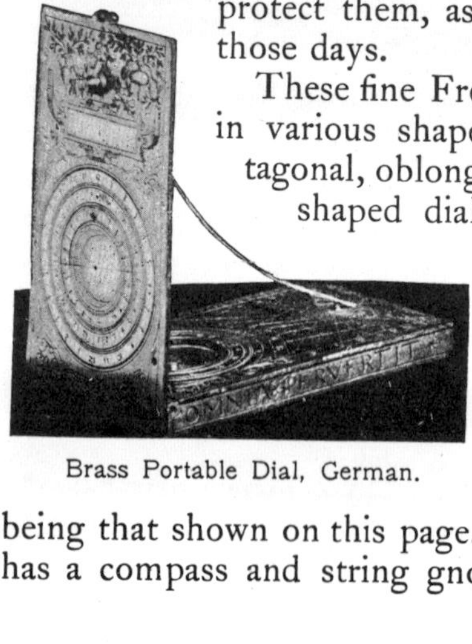

Brass Portable Dial, German.

At Nuremberg in the sixteenth and seventeenth centuries portable dials were made in great numbers, a favorite form being that shown on this page. This is of brass and has a compass and string gnomon, and elaborately engraved astronomical tables. It bears the motto *Omnia pervertit secli mutabilis ordo* 1571.

Other noted dial-makers were Alexius and Ulricus Schneip. On page 146 are two by these makers, and on this page one by V. S., dated 1572, probably Ulricus Schneip. These are all of gilt-brass, but vary decidedly in construction, and form quite a study. The one with raised standard with plummet is dated 1578, and is made by Alexius Schneip. The other on same page is dated 1553. It will be noted that all the many portable dials printed in this book have each some peculiarity of construction: no two are alike, many are unique, no others are known, and nearly all are here made public for the first time. They merit a fuller description, but any accurate accounts would fill a book rather than a chapter.

Brass Portable Dial, V. S., 1572.

An odd German dial (date about 1780) is given on page 130. The maker is D. Beringer. A hollow wooden block $2\frac{3}{4}$ inches square is supported on a standard. The lines and figures of the dial-face are printed on paper and pasted upon the block.

A brass octahedron is shown on page 129, a block with seven dials on its various faces. The horizontal dial on the top has no motto; the other mottoes read:—

OMNIA FIT ÆTAS, Time does all things.

ʽΟ ΗΛΙΟΣ ΠΑΝΤΑ ΜΕΡΕΙ, The sun divides all things.

FAC DUM TEMPUS OPUS, Work while it is day.

ΠΑΝΥ ΑΝΑΦΕΡΕΙ ΧΡΟΝΟΣ, The sun brings all things.

NULLA DIES SINE LINEA, No day without its line.

ʽΟ ΚΑΙΡΟΣ ΟΙΚΕΤΑΙ, The moment passes.

Such dials are very rare, especially in this size, about $7\frac{1}{2}$ inches in length. It is a portable dial, not a pocket-dial, as are most of these examples.

A very unusual block-dial is owned by Mr. Evans and shown on page 133. It was made in Styria,

Two Gilt-brass Portable Dials, 1578, 1553.

Austria, and has a decoration on one side of the arms of *Pälla Styria*.

Blocks with dials on several faces could be set in true position without the use of a compass; for if all these dials were accurately drawn for use in the same latitude, they would be in their true position when they all showed the same hour. Still more beautiful were the pocket-dials of ivory; with their delicate engraving and mellowed tint they seem an exquisite curio rather than a scientific instrument.

One now in the United States National Museum is shown on this page. It is an equinoctial pocket-dial, $3\frac{1}{4}$ inches by $4\frac{1}{2}$ inches in size. It has a compass, plummet, and string gnomon; and is elaborately engraved in colors with scales for length of days, lunar epacts, etc. The names of twenty

Ivory Portable Dial.

towns with the latitude of each are also engraved on this instrument.

On page 148 are two ivory folding-dials. The larger is a beautiful French dial of about 1660; it is octagonal in shape; $3\frac{1}{4}$ inches by $2\frac{1}{2}$ inches. Outside the cover there is an equinoctial dial. Inside are three silver discs: one is for using the dial by moonlight; the other two form a nocturnal dial.

Within the lid is the common string gnomon, dial, and compass, and silver calendar disc. Number 2 is the tiny folding-dial shaped like a book, — only $1\frac{1}{16}$ inches by $1\frac{3}{16}$ inches in size. It is ornamented with brass bosses. The pin-style dial shows the

Ivory Book-dial, 1630, and Octagonal Dial, 1660.

old German hours from sunrise. Its date is about 1630. On page 149 are three other ivory portable dials. The largest is a beautiful trinket as well as an object of use. It is oval in shape, $4\frac{3}{8}$ inches by $3\frac{1}{4}$ inches complete, with a chain to use as a chatelaine; I wonder what German *Frau* wore this pretty thing at her waist, and held it up and turned it round to the sunlight three hundred years ago. It was made by Hans Ducher in 1595, and is marked with his device, — a wriggling serpent between the initials H. and D. Hans Ducher or Tucher was a famous Nuremberg dialler who had a wonderfully pretty taste in dials and mottoes. We can say of him as an

Ivory Pocket-dials.

English lout of a farm boy said of Shakespeare. It was near Stratford, and a friend of mine spoke to the lad who was driving some conveyance for her through the country, wondering whether he knew of the great man, and what he knew. "He may ha' been very learned and all," drawled the boy, "but he didna know enough to spell his own name aright."

Ducher spelled his name with all the ingenious variety of which the sixteenth century was capable. This chatelaine dial was planned for ready reference: the points of the compass are engraved outside the lid, while the needle was visible through a tiny hole when the dial was closed. Around the face are these mottoes: *Hic marinus compassus semitam terra marique ostendit*; and, *Der spödter sol nichts verachten, den er kans besser machten.* On the bottom is a disc for using the dial by moonlight. On page 149 are two other ivory pocket-dials. One is an octagonal folding-dial $1\frac{7}{8}$ inches by $1\frac{7}{16}$ inches, marked *L. M.*, with a device, for Lienhart Miller, who manufactured dials from 1602 to 1643 in Nuremberg. The larger folding-dial is $3\frac{1}{4}$ inches by $1\frac{3}{4}$ inches; the work of Paulus Reinmann at Nuremberg, 1578. Two charming little ivory book-dials may be seen on this page. One is $3\frac{1}{4}$ inches by $2\frac{1}{4}$ inches; the other $2\frac{7}{8}$ inches by $2\frac{1}{4}$ inches. The first was made by Hans Troschel and has the fine motto *Hora fugit Mors venit.* The second is dated 1586 and has the motto *Sic transit Gloria Mundi.* The cover bears on one side the engraving of an armed man with a standard; on the other, a female saint holding a cross.

Ivory Book-dials.

Another form of portable dial is known as a quadrant dial. In them the principal circles of the sphere are projected upon a plane, instead of being indicated by rings or bands. The time is often shown by an adjustable bead which can be moved up and down upon a movable string,—a plumb-line. There were many shapes of these quadrant dials, some being of great quaintness and beauty, such as the "ship-dial"—made in the form of a turreted ship with a slider on the ship's mast and an ingenious arrangement for adjustment. A lyre-shaped horizontal and analemmatic dial belonging to Mr. Evans is pictured on this page. Its date is 1763. Another of unusual shape, with a movable gnomon, is given on page 135. This is the Thévenot dial. Hinged gnomons are found on many portable dials.

Lyre-shaped Horizontal Analemmatic Dial.

A very fine equinoctial universal dial owned by the author is dated 1764 and marked Filippo et Haves Fratelli di Bianchy. These dials were very costly, and were sometimes ordered by foreign governments for battleships or royal vessels. A similar one owned by Mr. Evans is given on page 152. This was made by Thomas Wright, Instrument-maker to the King (1720-1750).

Mr. Evans says of it, "It must have been a very costly instrument." I can say of my own dial, that it is one of the most exquisite pieces of brass work I have ever seen. The engraving is as beautiful as on the bridges of old verge-watches. These dials, and other of the costly dials have screws at the corners, by which they can be firmly attached to some fixed surface. A portable dial of special construction for latitudes 35°–90° is given on page 153. The maker was Augustin of Vienna. This is the only dial I ever saw of this form.

Dial made by Thomas Wright.

When metal workers were both artists and craftsmen, as they were two hundred years ago, hand-wrought articles varied more in construction and shape than to-day. None of the parts of a dial were turned out in vast numbers, as to-day of similar instruments; hence the variety, and hence the delight in collection.

A very odd and lovely portable dial was made at Nuremberg and perhaps elsewhere, what is called a goblet- or chalice-dial. This is a metal goblet marked within the bowl with the lines and numerals, while the gnomon is a perpendicular wire in the centre of the bowl. I had a great disappointment a few months ago in purchasing a metal dial dated 1624, which was described to me in so extraordinarily and minutely accurate a manner that I felt sure I had secured a chalice-dial; but upon opening the package a horizontal dial was seen in which the lines and numerals had been cut upon a pewter plate or shallow porringer, and the so-called date, 1624, on the base was, I fear, simply a tradesman's mark. Thus was added another to the many disappointments of a collector.

French Sun-dial.

> "As high as we had mounted in delight
> In our dejection did we sink as low."

A unique and beautiful casket-shaped dial belonging to Mr. Evans is shown on page 154. The base is $8\frac{1}{2}$ inches by 6 inches; the top 5 inches by $3\frac{1}{4}$ inches; the height 4 inches. The sides are silvered, the ornamental margins gilt. The corners of the bottom plate have four hemispherical recesses $1\frac{1}{4}$ inches in diameter, in which are the heads of four levelling screws. At the top and on the four slop-

Brass Equinoctial Dial, Augustin.

ing sides are five sun-dials, showing both Italian time (reckoned from sunset to sunset) and the ordinary hours. The gnomons of these dials are in the figure of a boy with outstretched finger. Each "time-boy" wears an encircling ribbon, on which is a motto. That on the top was missing, so Mr. Evans added an appropriate motto, namely, — *Opera manuum nostrarum dirige super nos.* Other mottoes read:

Casket Dial.

A solis ortu usque ad meridiem intervalla ipsa diei æque denuncio. A meridie usque ad solis occasum itaque cum illa gradior. Vespere cum eis pariter et mane in eodem die ostendere non deferam. Expecto donec veniat illuminatio mea ut cum aliis inservire valeam.

This casket was made in the year 1770; and I have noted a similar date on two or three Italian dials, of various makers, and I believe portable dials were much used at that time.

On one side is also the inscription, which can be read in the photograph. Within the lid is a hinged plate; on its outer face is a long inscription, giving the name of the owner and maker.

When the hinged plate is turned up, an inscription in Italian can be read, giving a full description of the use of the instrument. Within the casket are two cross-bars; one carries a compass, the other a plumb-level.

The introduction of the mariner's compass into Europe afforded a wonderful aid to the perfection and simplicity of the dial; and soon in the thirteenth century compass-dials appeared, and have continued in use to this day.

All portable dials of these general forms such as dials inscribed on quadrants, astrolabes, and flat plates of varied shapes, can be used without a compass and tell the time by the sun's altitude, not by its position. But more exact dials have a compass attached to them, and some have a spirit-level and a plummet; and some have set screws also. By these means a perfect adjustment can be made. Some have an ordinary dial-plate fitted with a quadrant and in such a way that the dial can be adjusted to suit the various latitudes.

It is said that all English soldiers have till this day the custom of taking a pocket sun-dial and compass as part of an outfit, since the latter will stand any climate and are not liable to the sudden derangements to which all watches are subject. Such are, I know, for sale in many London shops, but I think they are seldom found in America. Messrs.

Barker & Son make a large number of pocket sun-dials.

Finger rings were made with tiny compass and dial with a hinged cover. One was sold in New York two years ago, with an accompanying description saying that the gnomon was made of human hair which had grown on the head of "Bonnie Prince Charlie." It was suspected to be of very modern manufacture. There are several of these rings at the British Museum.

Brass Portable Dial. Maker, G. F. Brander.

A dial of unusual shape is in the National Museum at Washington; it is given on this page. The maker is G. F. Brander. Its construction is singular, but its action very accurate. A needlelike gnomon is mounted on a cross-bar, and casts its shadow on a semicircular sheet of brass beneath it.

In dialling the Chinese never accomplished anything, for they had scant and very deficient knowledge of astronomy and mathematics, though they had for forty-five centuries — according to the *Shu-king* — made instruments analogous to the quadrant and armillary sphere. Sun-dials came to them, and knowledge to make them, from the West, from the

Mohammedans, not through the Jesuits, as some writers have supposed. A few modern dials may be seen in public offices, all made under the direction of missionaries of the Roman Catholic Church.

From remote antiquity the manufacture of pocket compasses has been held exclusively by a family

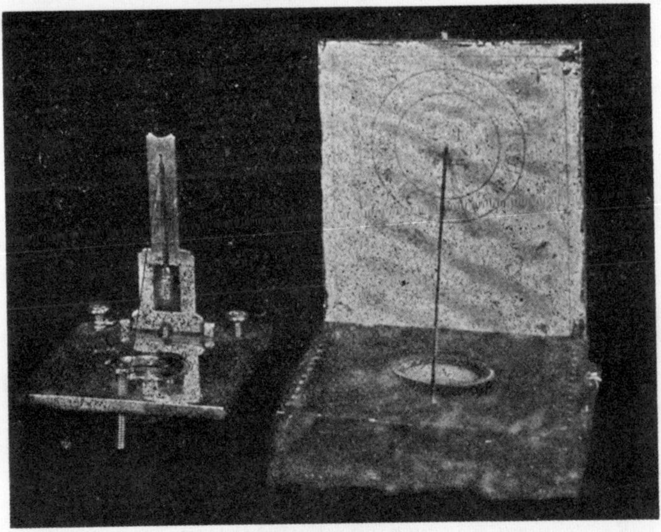

Chinese Sun-dials.

named Wang living in Hiri-wing-hien, in the province of Nyanhwai. Many of these compasses have a sun-dial attached; these are in the German type, with a thread for a gnomon. Another portable dial used by clock-makers is marked with notches, one for each month, thus giving the gnomon varied angles. The pocket-dial of the Japanese exceeds that of the Chinese in every respect.

Two Chinese dials are shown on page 157. Number 1 is a little Chinese folding-dial and compass, in a wooden case. It was bought by Mr. Evans in a shop in Chinatown, San Francisco, 1877. The second is a Chinese dial with compass and plummet. On this page is shown a circular silver box containing a compass and dial. This is Japanese.

There are many fanciful "conceits" for portable dials. The human hand is made into one by holding it upright and thrusting a small stick between the thumb and forefinger, and turning the loose end about until the shadow of this gnomon falls on "the line of life." I make also an amusing little toy sun-dial, using as materials a plain glass cylinder-shaped tumbler, a knitting needle, and a slip of cardboard. This is a surprisingly accurate timekeeper.

Japanese Pocket-dial.

Some very fine and exact portable dials are manufactured to-day. Flechet's universal sun-dial, shown on page 159, is of great value to a traveller or explorer. It will be seen by any student of dials to be a general gnomonic form which has long been made, but in addition the curve of the mean time has been traced upon the circle *H* in such a manner that by an adjustment of the horary circle and an inclination of the axis, the latitude of the plane can be determined; then of course the instrument can be adjusted to give the exact time of that place.

A very interesting dial is sold in London which was devised by General Oliver, and the portable form is known by his name. It indicates mean time or clock time, which may differ from apparent time as much as sixteen minutes. This time-adjustment is accomplished simply by an enlargement of a section of the rod which is the gnomon. Thus a curved shadow is thrown instead of a straight one. The meridional and equatorial arcs are adjustable, and have screws for fixing them at any given altitude. The shadow of this perfected gnomon falls on the equatorial circle. This dial and gnomon are an adaptation of the principle used in the cannon-dial shown on page 49.

Flechet's Universal Sun-dial.

An ingenious card-dial dates from the early part of the seventeenth century. It was sometimes called The Capuchin from a fancied resemblance to a Capuchin cowl thrown back. A noted mathematician says of it: —

"This card-dial deserves to be looked upon as something more than a mere toy. Its ingenuity and scientific accuracy give it an educational value which is not to be measured by the roughness of the results obtained."

This card-dial, calculated for the latitude of London, was patented by James Ferguson in 1759, and apparently was sold in large numbers. It was on a card $4\frac{1}{4}$ inches by 3 inches, with a tiny green silk cord as a marker. This could be carried in a pocketbook. It has been republished in this country, for the latitude of New York, by a true "dialler," Captain John S. Bailey of Buckingham, Pennsylvania. I give a figure of his card on page 161.

German Portable Dial, Brass. Owned by Lewis Evans, Esq.

Of course, there are in the United States many houses that manufacture optical and mathematical instruments and also make sun-dials. There are also those who make and sell very pretty brass dials, made to look well, and absolutely regardless of the shape of the gnomon or drawing of the hour lines. I know no individual, however, save Captain Bailey, who makes accurate sun-dials for all latitudes. His picture, taken in his workshop, is shown opposite page 162. He was originally a house-builder, then a clock seller and mender, then gradually drifted into dial-making. He has made and sold over three

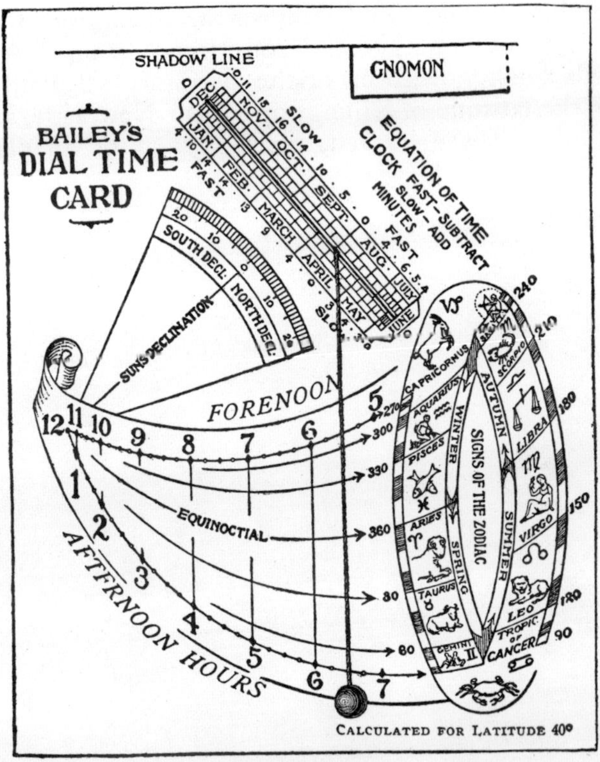

Bailey's Time Dial.

hundred dials; some of them have been very handsome and costly; his simple metal, horizontal dials have had a steady sale. His books of instruction were Leadbetter's *Dialling* and Ferguson's *Lectures*, the same book from which George and Robert Stephenson, the great railroad engineers, constructed the vertical dial which may still be seen on the humble cottage at Killingworth, which was their early home. Father and son learned from Ferguson's book how to make the necessary calculations for the latitude of Killingworth, near Newcastle, and together hewed, carved, polished, and set in place the dial-stone.

John S. Bailey, Dial-maker, in his Workshop.

CHAPTER VII

THE SUN-DIAL AS AN EMBLEM

"An Emblem is but a silent Parable. Before the knowledge of letters, God was known by hieroglyphics. And what indeed are the Heavens, the Earth, nay every creature, but Hieroglyphics and Emblems of His Glory? I have no more to say: I wish thee as much pleasure in the reading as I had in the writing. Farewell! Reader!"

—*Emblems Divine and Moral* (*The Preface*), FRANCIS QUARLES, 1634.

"How the face of common day
Is written all over with tender histories,
When you study it that intenser way.
Till now you dreamed not what could be done
With a bit of rock, and a ray of sun."

—JAMES RUSSELL LOWELL.

NEARLY all original forms of sun-dials were good, because simple and conventional, and certain decorative ideas which appeared after a time were also good; but sun-dials in general, as made to-day, have scant originality. This is surprising when we note the originality of design displayed of late in architectural decoration. I believe the true reason for this monotony is the fact that the sun-dial is looked upon solely as an architectural detail, not as an individual object with a place of its own in the world; there is a lack of appreciation of the ideality of the sun-dial. All this commonplaceness of design will vanish, all this monotony cease, if designers learn to love the sun-dial, and to regard it as an Emblem, and one of importance, an Emblem with a capital initial, Emblem in the highest, the specific, the Shakespearian sense of the word, Emblem with a moral lesson to suggest or even tell in detail.

Let me explain precisely what I term an Emblem. The word has had many meanings, and offers a curious example of word-variation. Originally its classical meaning was simply an inserted ornament which had some significant place—often this was of mosaic or marquetry work; then it became a detachable ornament, still significant, which was hung on pillars or vases; then it became simply the "picture and short posie expressing some particular conceit" which is the definition in Cotgrave's Dictionary, which I fancy was the dictionary known of Shakespeare.

A noted Emblem writer, Geoffrey Whitney, gave in the year 1586 a lucid explanation of an Emblem in such definite, albeit quaint, phrasing that I quote it in full.

"This worde Embleme is as much to say in English as To set in or to put in: properlie ment by suche figures or workes as are wrought in plate or in stone in the Pavementes, or on the Waules, or suchlike, for the adorning of the place: having some wittie Devise expressed with cunning woorkmanship, something obscure to be perceived at

Post amara dulcia.

TO M THOMAS MYNORS.

SHARPE prickes preserue the Rose, on euerie parte,
That who in haste to pull the same intendes,
Is like to pricke his fingers, till they smarte?
But being gotte, it makes him straight amendes
It is so freshe, and pleasant to the smell,
Thoughe he was prick'd, he thinkes he ventur'd well.
And he that faine woulde get the gallant rose,
And will not reache, for feare his fingers bleede;
A nettle, is more fitter for his nose?
Or hemblocke meete his appetite to feede?
None merites sweete, who tasted not the sower,
Who feares to climbe, deserues no fruicte, nor flower.
Which showes, we shoulde not fainte for anie paine,
For to atchieue the fruictes of our desire:
But still proceede, and hope at lengthe to gaine,
The thinges wee wishe, and craue with hartes entire:
Which all our toile, and labour, shal requite,
For after paine, comes pleasure, and delighte.
When winter endes, comes in the pleasant springe.
When nighte is done, the gladsome daye appeares.
When greifes be gone, then ioye doth make vs singe.

Claud. in nuptiis honosij
Non quisquam fruitur veris odoribus,
Hyblaeis latebris nec spoliat fauos,
Si fronti caueat, si timeat rubos.
Armat spina rosas, mella tegunt apes.

Dulcia non meruit qui non gustauit [illegible]

Page from the *Emblems* of Geoffrey Whitney.

the first, whereby when with further Consideration it is understoode, it maie the greater delighte the beholder. And although the Worde doth comprehende manie things, and diuers matters maie therein be contained, yet: all Emblems maie for the most parte, be reduced into those three kindes, which is *Historicall*, *Naturall* and *Morall*. *Historicall* as representing the actes of some noble persones, being matter of Historie: *Naturall* as in expressing the nature of creatures, for example the Loue of yonge Storkes to the ould, and suchelike: *Morall*, pertaining to vertue and instruction of Life."

This careful explanation illumines the meaning I wish to convey of the sun-dial as an Emblem: the dial should, like any other Emblem, have "some witty device," be "of cunning workmanship," have "something obscure not to be perceived at first," and should "greatly delight the beholder." And the significance of the sun-dial could also be classed very clearly under the heads, historical, natural, and moral. Another sentence of Whitney, in the title-page of his book, *A Choice of Emblems*, 1586, runs thus:—

"A Worke adorned with varietie of matter, both pleasant and profitable, wherein those that please may finde to fit their fancies. Herein by the office of the eie and the eare, the minde may reap double delighte through wholesome preceptes, shadowed with pleasant devises: both fit for the vertuous, to their incoraging: and for the wicked, for their admonishing, and amendment."

Which words should also apply to the sun-dial: Let it be literature to the bookless; a monitor to the heedless; an encouragement to the virtuous; a reproof to the wicked; let it be a delight to the eye, and above all let it be significant of *something*, be this historical, natural, or moral. It must have some implied meaning in addition to its actual presentation or it is not an Emblem.

Emblems are ancient devices. Dutch, German, and Italian authors wrote Emblems in profusion by the close of the fifteenth century. Then these books suddenly multiplied in all European languages, and with the improved art of pictorial illustration reached a high degree of excellence. Our great English authors, Spenser and Shakespeare, show plainly the influence of Emblem books. A splendid book called *Shakespere and the Emblem Writers*, by Henry Green, reveals the similarity of thought and expression shown by them all; it also tells of all the chief Emblem books, scores in number, previous to the year 1616, A.D. The general conception held of a book of Emblems is of a child's book, and it is true that they are most appropriate for that purpose. It is natural for children to like Emblems, and they understand them. Emblems suit their fancy. Stevenson says, "Making believe is the gist of the whole life of a child. Children are content to forego what we call the realities and prefer the shadow to the substance." Others think of a book of Emblems as having always a spiritual or religious meaning, such as the *Emblems* of Francis Quarles or Willet's *Century of Sacred Emblems*. In truth an emblem should be serious; it does not welcome flippancy any more than does a sun-dial. But it is not a sacred device, nor even a religious one. A study of the old emblematic books such as *The Dance of Death*; *The Ship of Fools*, 1500; *The Dialogue of Creatures*, 1530; *A Garden of Heroicall Devices*, 1612; *A Display of Heraldry*, 1611; *Allciat's Emblems*, 1549; Holbein's *Imagines Mortis*, 1545, would doubtless give to a ready mind frequent suggestion for sun-dial design.

Sun-dial at Prestbury, England.

A great love of Emblems sprung up in France through the wonderful popularity of the poet Theodore Beza, whose story is so pleasant to read. In France, Mary, Queen of Scots, learned to love Emblem books, as did her son James. And through this love grew his great interest in the allegorical representations dubbed masques, which so prevailed at his court after he became king of England. I will tell at some length of Mary's interest in Emblems, as the story affords a striking example of the part they played in the history of that day.

The many Emblem books furnished to her, as to other English women, beautiful and abundant designs for the decoration of houses and furniture. In Drummond's *History of Scotland* is a letter to Ben Jonson, from which we learn of a wonderful piece of needlework of Mary, Queen of Scots; the letter begins thus: —

> "I have been curious to find out for you the Impresses and Emblemes on a Bed of State wrought out and embroidered all with gold and silk by the late Queen Mary, mother to our sacred Sovereign: the first is the Loadstone turning towards the pole, the word her Majesties' name turned on an Anagram. Marie Steuart sa virtu, m'attire. This hath reference to a Crucifix before which with all her Royall ornaments she is humbled on her knees most lively."

There was also an imprese of a phœnix in flames; an Apple tree growing on a Thorn; an arrow passing through three birds; Caduceus with two flutes, and a peacock; two women upon the wheel of fortune, one of these (a figure of peace with a cornucopia) signified Queen Elizabeth; a pyramid overgrown with Ivy; a ship with her mast broken and fallen in the sea; a big lion and whelp; a lion in a net with hares passing wantonly over him; a Palm tree; a bird in a cage with a hawk flying about, and the motto "'Tis ill with me now, but I fear worse betides me"; a triangle with the sun in a circle; a porcupine among rocks; a portcullis; an "impresae of Henry VIII"; one of the Duke of Surrey; the Annunciation; a tree planted in a churchyard environed with dead men's bones; eclipses of the sun and moon; a sword cast in to weigh down gold; a Pine tree watered with wine; a wheel rolled from the mountain to the sea; a heap of wings and feathers; a "Trophie upon a Tree with Mytres, Crowns, Hats,

Masks, Swords, Books, and a Woman with a Vail about her eyes or muffled, pointing to some about her."

One of the noblest of these allegories gave three crowns, two below and one above in the sky. The motto was "And awaits another," implying that her crowns of France and Scotland would be ended with a crown in heaven. A homely but most significant design showed the plant Camomile growing in a garden; the motto, *Fructus calcata dat amplos*, — Trampled upon, she gives out greater fragrance. A panoply of war, composed of helmets, lances, pikes, muskets, cannon, had the motto, *Dabit Deus his quoque finem*, — God can put an end to these things.

Sun-dial at Adlington Hall, Cheshire, England.

The full royal arms of England, Scotland, and France "severally by themselves and all quartered" appeared in every part of the bed. It was said that "the workmanship was neatly done and above all value." Nearly all the designs bore a Latin motto, also the name and title of the person or event in politics or history signified by each of these twenty-nine Emblems. The designs for these embroidered Emblems were taken from various books, from Whitney, and from Jovio's *Design of Love;* one of the latter was "the Impressa of King Francis First, a salamander, which signified that he was burning with love for the Queen and sought the flames."

This wonderful piece of needlework might well have been the one praised by the Water Poet in 1640 in these lines: —

> "Flowers, Plants and Fishes, Beasts, Birds, Flyes and Bees
> Hills, Dales, Plains, Pastures, Skies, Seas, Rivers, Trees
> There's nothing near at hand, or farthest sought
> But with the needle may be shap'd and wrought;
> Moreover, Posies rare, and Anagrams,
> Signifique searching Sentences from Names,
> True History or various pleasant Fiction
> In sundry colors mixt, with art's commixion;
> All in Dimension Ovals, Squares and Rounds
> Art's life included within Nature's bounds."

This extraordinary bed, so "curiously wrought," has wholly disappeared; we might doubt the possibility of any one bedstead holding all these designs were it not that at Hinckley in Leicester there is another having the same number of emblematical designs and "Latin mottoes in Capital letters conspicuously introduced."

I have no lovelier picture in my mind of this fair queen — and she is often pictured before me in my day-dreams — than her presence as she sat bending over her embroidery frame, needle and crewels in hand, steadily working upon the marvellous covering of this great bed; working through the long weary hours of the dark winter days; working at the deep-recessed, ill-lighted windows of the thick-walled Scottish castles which were her prisons; working at the slit-like gun-windows of her later dungeons; working by the scant firelight so grudgingly supplied her; working by the dim and tiny cruisie of her day, or by waxen tapers; and often working with that wonderful cheerfulness which seems to have been God-given to her. She found, I trust, the comfort which every good needlewoman has in doing good needlework.

> "Yet howsoever Sorrow came or went
> She made the needle her companion still
> And in that exercise her time she spent."

She is not the only woman who has turned to her needle as the only thing which could occupy and comfort her grief-filled days. I have wondered whether in the many thoughts that crowded her ever active brain, she had no illuminations of the future, whether she did not thus work with the thought, the hope, that through this needlework she could send a message to succeeding centuries, that women, certainly, would understand.

Sun-dial in Inner Temple Garden, London.

It was certainly natural that this unhappy creature should turn to symbols and devices, to allegories and prophecies, with a despairing hope of a happy end to all her troubles. With her superstitious nature we can easily believe that in those symbols she both rejoiced and trembled. Her own personal devices were many and varied; all were interesting. After the death of her boy-husband, the dauphin of France, her device was a Liquorice plant; the root only of this is sweet, and that is underground. Her motto was *Dulce meum terra tegit*,—The earth covers my sweet one. Another was a vine from which the withered branches are being pruned by a hand with a pruning-bill. A third was an Apple tree growing on a thorn; the motto, *Per vincula crescit.*

In the family archives of the Earl of Leven is a letter written by her in which she orders embroidery materials just as we might have done yesterday.

> "Ye shall not fail to send with this bearer to me a half-ell of *incarnatt* Satin, and a half-ell *blew* Satin; also more twined silk gif there rests any, and sewing silver and sewing gold; . . . with twa ounce black sewing silk. . . . Ye shall cause make ane dozen of raising needles and moulds and send me. And speir at Sewals gif he has any other covering of beds to me nor green."

After she received these materials she worked for many months upon a magnificent over-garment for Queen Elizabeth, with a significant design; and showed true womanly pleasure when it was finished and despatched with a letter to the queen, whose

Obelisk-shaped Dial in Garden at Linburn House, Midlothian, Scotland.

faded tawny hair would ill consort with the carnation satin when compared with the darker locks of the Scottish queen, who, I doubt not, "tried it on" again and again in process of making.

Mary received from another source four hanks of gold thread and moulds and needles for "raising," which was the heavily embossed and difficult embroidery much in vogue in her day. A splendid screen of her "raised" work still exists, and is most valuable as a record of the costumes of her day; every detail is given; the jewellery is worked in satin-stitch with glazed yellow flaxen thread and the pearls are tiny knots. Fardingales, ruffs, and fans, all are faithfully depicted. The drawing of the figures is animated and good. Sir Walter Scott thought it represented some old French or Italian ballad or romance; while Miss Strickland thought it pictured the ill-fated love of Mary and Darnley, as the gallant knight of the screen seems to spring from two Marguerite flowers, which apparently typified the two queens, Margaret Tudor and Margaret

Sun-dial in Churchyard, Rostherne, Cheshire.

Lennox, from whom Darnley claimed descent. The woman's figure wears the costume of Mary, and bears the Rose cognizance. One singular and inexplicable panel shows a gentleman seated with his leg bound and stretched on a block and about to be cut off, apparently by the order of a stern queen who stands near, while other ladies of the court turn away in horror. Taylor might have written of her instead of Katherine, wife of Henry VIII: —

> "Although a Queen, yet she her days did passe
> In working with the Needle curiously,
> As in the Towre, and places more beside,
> Her excellent memorials may be seen
> Whereby the Needle's prayse is dignified
> By her fair ladies, and herself, a Queen."

The tenderness which most women have for the history of this unhappy queen comes largely through her womanliness. We are drawn to her through her instinct in womanly doings. She took great pleasure in gardening, filling the gardens at Holyrood with flowers and trees from France. Two beautiful Plane trees stood till this century, and were pointed out as her trees. Her sun-dial was removed to Fingask Castle in Perthshire. Nothing could seem so close to her daily life as this sun-dial. I know not whether it still exists, nor what it was like; but there were beautiful dials in Scotland in her day.

Other English women adorned their closets with embroidered emblems; forty-two were in the painted closet of Lady Drury at Hemstead. Samplers were found in every household, the work of every housewife, of every woman-child who reached the age of ten, and of some not more than half that age. Often they have an imprese or a family crest. They form a fascinating example of domestic Emblems. Whatever women could do with their needles served to perpetuate Emblems, for their imaginative side appealed to a woman's nature; often women ordered the erection of symbolistic pillars and sun-dials.

The original Emblem, any figure or ornament made for a sun-dial by sculptor, painter, engraver, or architect, by any worker in stone, metal, or wood, should be symbolized; it should be the sign or token of a saying, an event, a thought, a sentiment, a fancy, a quality of the mind or heart, a peculiarity or attribute of character, any abstract idea,—nay more, it might mark an operation of the soul. The devices of heraldry can aid greatly in giving the history of a man, a family, a race, a nation. Many crests and coats of arms are truly emblematical; and therefore heraldry offers an infinite variety of suggestions for sun-dials. There were heraldic devices on sun-dials, and sun-dials were used as devices in heraldry. The Emblem here given is of special interest to us because the device is a sun-dial. It was the favorite Emblem of the gentle and neglected wife of Henry III of France,

Emblem of Louise de Valdemont.

Louise of Valdemont. Above the sun-dial, which is on a pedestal, shines the full-rayed sun. On a ribbon the meek motto, *Aspice ut aspiciar*, — Look on me that I may be looked on.

Sir Philip Sidney also had a sun-dial as his personal emblem, and it was chosen in order "to acknowledge his essence to be in his gracious Sovereign" — whatever that may mean. It was a sun-dial with the sun setting; the motto, *Occasu desines esse*. Relying upon his prince's favor, he devised the sun shining upon a bush, inscribed *Si deferis pireo*. To indicate the persistency of his character, he had another Emblem representing the Caspian Sea, which never ebbs nor flows, and the motto *Sine Reflexa*. Another rather sacrilegious device showed his love for his fair lady; a Venus in a cloud with the motto *Salve Me Domina*. He had several other impresses to signify courage, assiduity, and also revenge.

Frederick Cornaro, Bishop of Padua, had as a device a Rose with this sun-dial motto: *Una dies aperit, conficit una dies*, — One day opens, one day ends it. This seems to me perhaps a bit fanciful for a sun-dial, save for one designed especially with the thought of the life of a day — such as one with a floral design.

Another personal emblem or device existed from mediæval days, and was known as a badge. A crest differed from a badge in that the former was worn only on the helmet and by its proper owner, while the badge was worn by followers or retainers, and was placed on the sleeve or breast of the body garment. These formal badges or cognizants often alluded to a name, an estate, an office of honor, or some family exploit, a deed of valor or of rare happening, or an escape from death; they glittered on standards, were embroidered on the dress on sleeve or breast, or when fashioned in metal were worn on the sleeve. These were hereditary, and a few are

Washington Sun-dial.

still retained in old English families; among them are the Stafford knot and the Pelham buckle.

The introduction of a coat-of-arms or crest upon a dial gives it at once an emblematic value, and still more, a personal value.

On this page is shown an ancient dial-face, which is of much interest to all patriotic Americans, because it bears a dated Washington coat-of-arms, which is believed to have furnished the notion of our national

Ancient Cross at Great Brington, Northamptonshire.

"Stars and Stripes." This dial was found at "The Washington House," Little Brington, Northamptonshire, England. It is a circular slab of sandstone sixteen inches in diameter. The letters R. W. can be dimly seen. These were probably the initials of Robert Washington. On page 178 is the old cross just outside the churchyard at Great Brington. This may once have held a dial. Its date is earlier than 1400. The ancestors of George Washington must have passed close to this cross every time they attended church. In the yard of the rectory of this Great Brington church stands the dial shown on page 181; this is at the home of A. L. Y. Morley, Esq., the faithful antiquary who has given me the many sun-dials from Northamptonshire which are shown in this book. The motto on his dial is most quaint.

> "Haste! oh Haste! Thou Sluggard, Haste!
> The Present is already past."

It was natural that in the highest forms of Emblem making and emblematic writing color should take an important part; it did so directly and also in some occult ways, of which we have had a hint for many a year. In 1886 Arthur Rembaults put this hint into verse in his sonnet about the colors of vowels. Novelty-seeking French folk eagerly queried to other French folk, as though asking a conundrum, "What colors are your vowels?"

> "A, black; E, white; I, blue; O, red; U, yellow;
> But purple seeks in vain its vowel-fellow,"

wrote the poet.

Walter Savage Landor had such a profound sense of color that he had a language or standard for colors. Purple expressed grandeur of thought;

scarlet, vigor of expression; pink, liveliness; green, an equable composition. I have long had another notion, — that I should like to use vari-colored printer's inks, printing certain words in certain colors, or using a specially symbolic tint for a certain chapter.

We can remember the value of symbolic color when painting sun-dials; as, for instance, those on the wall of a house. In continental Europe painted sun-dials are seen constantly, even on very humble houses, and most effective and interesting they are; they might well be adopted on country houses in America. For handsomer mansions, when chosen with thought and taste and fitted to the style of the architecture of the house, a painted and gilded dial-face has many advantages over a carved one. It can be seen more distinctly if upon the high wall of a house, and can readily be kept in freshness. I saw recently upon a half timbered house, on the end of a gable, a painted sun-dial in heraldic colors which seemed to me the perfection of good taste. Upon the long stables of an English country house, where the original timbering and external beams of the early barn structure have been carefully carried out, there is a richly painted sun-dial facing each point of the compass, so that the time of the day can readily be told on all sides by farm and stable workers.

There is no doubt that any object or any deed which has or has had a symbolic meaning receives through this a certain charm, a charm occult and often scarcely formulated, yet nevertheless present.

This subtle interest exists in very commonplace objects; we feel it in sign-boards, in sign-posts or guide-boards; let us see why. Whence is the word *sign?* Think of the very word, and you have the key to the secret — and to the interest. I never wearied in weeks of research about sign-boards, hanging signs, for my book, *Stage-Coach and Tavern Days*, simply because they had that inexplicable charm. I never cease to feel a half-liking for guide-boards, which are fast disappearing because useless in our days of travel by railroads and electric cars. You find them now at the angles of the road, flat in the grass and bushes, or standing twisted askew, pointing skyward or nowhere. Their very place at the cross-roads has a significance, what or whence I know not, for it is a significance of "forbidden." We know that in our country even in the nineteenth century suicides were buried at the cross-roads; buried in the years preceding, with a stake through

Sun-dial in Rectory Garden, Great Brington, Northamptonshire.

their hearts — a cruel old Dutch and English law. Judge Sewall tells of the public obloquy and horror of suicide in Massachusetts in colonial days. Under a heap of stones suicides were buried at night in deepest disgrace; one at a Connecticut cross-roads thus slept in ignominy till it was discovered that the poor fellow had been murdered.

A more grotesque sight still saw the old New England cross-roads in a "shift-marriage," when a widow, "clad only in her shift," was thus married to her second husband. By this ordeal she was freed from liability for her first husband's debts. This was an ancient symbol, derived from the marriage investiture of the Orientals. In Dutch New York a widow obtained future immunity from debt by placing a key or a straw on her dead husband's coffin and then taking it away. The use of a straw or turf as a symbol to indicate worldly possessions extended to legal transfers — when real estate was conveyed by livery of seisin. A new owner was given corporal possession by transferring to him, if the property were a house, a key or door-latch; if land, a turf or twig. It formerly accompanied all transfers. The word *livery* in its many symbolisms forms an interesting word-study; the meaning of giving possession, and in sequence, the delivery itself. Liveries were allowances of food, clothing, and other provisions, as in the army, or to a great family; from this, keeping on a regular allowance as applied to horses only, — as to keep a horse at livery; also a second meaning of an allowance of uniform garments, and hence a regular dress for servants.

A weather-vane is another everyday object with a halo of interest as a symbol; the cock was the natural herald of the day, and the weathercock now employed to show "the way of the wind," was originally a sun emblem. In the symbolic writing of the Chinese, the sun is represented by a cock in a circle. Beautiful ancient gems exist, — some are in the possession of ancient societies of Free Masons, — cut with the figure of a cock, meaning the sun.

All symbolists would of course have us include in these objects of mysterious influence all architectural erections conspicuous for height and slenderness, such as obelisks, steeples, minarets, tall towers, and upright stones and monuments, under the assertion that they represent the pyramidal forms of fire, and have had a symbolic meaning ever since the days of the fire-worshippers. Certainly we will not deny that they have a strong influence; the tall steeple of the New England meeting-house doubtless had an earlier form in the "reminding-stone," the monuments of earlier days. The Bible records the setting up of monuments by the patriarchs. Monoliths are known of in all early religions. In the turreted temples of the Bhudds, in the fire-towers of the Sikks, in the spires of the Hindoos, in the flame-fanes of the Parsees, in the pyramids of the Egyptians, we find testimony to the deification of fire. Many of these pyramidal forms bear emblems of sun-worship; some bear sun-dials; many are the gnomons of sun-dials.

In writing of Emblems we have to resist firmly constant inclination to turn into side paths and walk therein gossiping garrulously; one of these by-paths would lead us to write at length of the symbolical language of mythology and of ancient art. As specially allied to the subject of sun-dials, we are led to learn all sun-lore; to know of sun-worship, of sun-tradition, of sun-influences in various sciences, especially medicine. The mystic doctrines of ancient Greece are not wholly lost in daily life to-day. These doctrines were conveyed by allegories and symbols which had a character of sanctity. Many of the emblems of these mystical religions are found now in our Christian churches. Such are the rose-window, the altar and candles.

CHAPTER VIII

SYMBOLIC DESIGNS FOR SUN-DIALS

"How beautiful your presence, how benign,
Servants of God, who not a thought will share
With the vain world; who outwardly as bare
As winter trees, yield no fallacious sign."
— *Ecclesiastical Sketches*, XIX, WILLIAM WORDSWORTH.

THE old Emblem writer, Geoffrey Whitney, noted with severity, as we have with sympathy, that readers and on-lookers weary of monotony of device: "The nature of Man is alwaies delighted in nouelties & too muche corrupte with curiousnes and newfanglenes." Truly we desire and need "curiousnes and newfanglenes" in sun-dials as in all things else, and to satisfy that desire I shall offer in this chapter some suggestions for novelty as well as significance of design in sun-dials.

At Ophir Farm, the country seat of Hon. Whitelaw Reid, there stands in an open court, near the house, a sun-dial. It is pictured with a corner of the castle-like house opposite this page. The dial is set upon a circle of brick pavement surrounded by sentinel trees of Japanese Retinosporas. It is not in a garden, but stands rather sombrely alone, with no flowers, no creeping vines, no neighbors but the solemn trees. It is fitting that it should thus stand, for it is an emblematic dial, and was not meant to be lightly wreathed and garlanded, nor to have its significance hidden. The dial-plate rests upon a carefully wrought bronze tortoise, and that is supported on a symmetrical marble pillar which bear designs of the signs of the zodiac in wrought bronze.

The design of a tortoise is most appropriate for a sun-dial. The myth of the tortoise is world-wide. The Hindoos believe that a great tortoise lies beneath the earth on his back. Earlier is the notion that the earth itself is a tortoise; the flat plate on the belly of the tortoise is the land, and the sky is the shell of the back.

The oldest of all Chinese pre-Confucian books is the famous *Book of Changes*. It contains a system of philosophy deduced from eight hexagrams which were copied from the lines on the back of a tortoise. Each represents some great power in nature, as fire, water, earth, etc. It is also regarded as a calendar of the lunar year. So important are the lessons of this book, so great is its wisdom, that Confucius declared that could a hundred years be added to his life he would devote them all to its study.

A most suitable and perhaps the most dignified engraving for a dial-face is a chart showing the lines of latitude, signs of the zodiac, etc. Such a face is,

Sun-dial at Ophir Farm, near White Plains, New York. Seat of Hon. Whitelaw Reid.

of course, a costly one, as the drawing must be made with astronomical precision and copied by a skilled workman. I give on page 187 such a dial-face from

Dial-face with Lines of the Zodiac. Owned by Author.

my collection. It is about ten inches square, of finest workmanship, dated 1812; a most attractive piece of work. The dial on the church porch at

Eyam (facing page 64) shows the lines of the tropics and the equinoctial.

On this page is given Captain Bailey's fine dial, which tells the seasons. This dial exhibits beautifully, not only the comparative obliquity and directness of the sun's light, which is the primary cause of the seasons, but it also shows the rapidity of what in nautical terms is called the sun's declination and entry into the signs of the zodiac. As the sun gets higher, the dial's shadow goes down; then it crawls from Cancer up through Leo and Virgo to Libra, and so on.

Captain Bailey's Seasons'-dial.

The signs of the zodiac are shown in bronze on the pedestal of Mr. Reid's dial. Being an absolute symbol of the progress of time, they are a natural and beautiful emblem for the decoration of pedestal as well as dial-face, and have appeared on many. Save in the simplest symbol they cannot become a common decoration, as they would be too costly a one, whether cast in metal or carved in wood or stone. There is a wide range of decorative forms to choose from, and many of them quaint indeed. They ever have had a fascination for me since my childhood, when I gazed with bewildered curiosity upon their representation in old almanacs. From almanacs more ancient than those of my youthful days, many hints may be obtained for sun-dial designs. We can scarcely go back to Babylon where, 2000 B.C., the zodiac was formed; but one of the oldest drawings of the zodiac is in an astronomical manuscript of the fourteenth century, in the Chetham Library. Each month has a picture medallion as a device, and each line of the following verses is explanatory of the device of month:—

> "Over yis fire I warme myn handes.
> Wyth yis spade I delve myn landes.
> Here knitte I my vynes in springe.
> So merie I hear yese foulis singe.
> I am as joly as bird on bouz,
> Here wede I corn, clene I houz.
> Wyth yis sythe my medis I mowe.
> Here repe I myn corn so lowe.
> Wyth ys flayl I yresche my bred.
> Here sowe I my whete so red.
> Wyth ys knyf I styk my swyne.
> Welcome Christmasse wyth Ale and Wyn."

These spirited verses have a real Chaucerian ring; and it amuses me to see that the spring house-cleaning is not a Yankee invention. This old manuscript contains an astrological volvelle, an instrument mentioned by Chaucer. It has seemed strange to me to be able to buy within a year one of these astrological volvelles, made, I am sure, from ancient patterns in evidence since Chaucer's day. This old manuscript must be the original *Farmer's Almanac*; and the French *Kalendrier des Bergers* is equally longeval. I may be permitted to gossip a bit about almanacs since they have a cousinship with sun-dials.

Zodiacal symbolism was conspicuous in mediæval art. Nearly all the French cathedrals of the twelfth to fourteenth centuries have the signs of the zodiac painted or cut on their gates. Of course the zodiac of Notre Dame in Paris is the best known. Several English churches have zodiacal decoration, some being very elaborate. In Padua there was a curious sun-dial of the months, wherein the sun's rays struck in turn the twelve symbols, which were painted by Giotto. The superb dial on the Church of Our Lady at Munich, with the signs of the zodiac, is shown in this book. Our modern hieroglyphics representing the signs of the zodiac were known in the tenth century. The signs of the zodiac are four thousand years old. Those of the Oriental lands—as for instance the zodiacal signs of India—are beautiful and decorative to a degree. Let us remain on our own continent, and take the calendar of the Aztecs, and we at once have a wonderful field of beauty, variety, and suggestion. The "age" of the Mexican calendar was fifty-two years; this was composed of four cycles, each of thirteen years. The single year was named significantly a word meaning new grass. The cycles were designated as the flint, rabbit, cane, and house. Nothing could be more beautiful or appropriate for a dial-face than the Mexican representation of a cycle as shown on page 191. The outer edge was painted with an encircling snake holding the tip of its tail in its mouth and having a twist or knot in its body at each of the four cardinal points; within was a close border of the four cycle symbols each recurring thirteen times. Within this were the signs of the month.

The Aztec symbol for the year and the month is equally beautiful and appropriate; the former having in the centre their emblem of the sun, with a striking border of the designs for the eighteen months; while the Aztec month was divided in similar manner into days. These are within the compass of any skilled designer and workman, either in brass-engraving or stone-cutting. A truly glorious sun-dial, one of profound historic association and of exquisite fitness for the purpose, might be made by adopting the design of the Aztec calendar-stone dug up in the city of Mexico a hundred years ago, and so eloquently described and explained by Gama, see page 193. This great sun stone is about nine feet in diameter. In the centre is a drawing of the sun as usually painted by the Mexicans, and this is surrounded by beautiful hieroglyphics. Of course the careful workmanship of this remarkable carving

Aztec Calendar.

need not be copied; but the suggestive outlines and general shape could be followed, and with wonderful effect. As every line of this Mexican dial had a sun-meaning, its appropriateness would equal its elegance. The Peruvians had similar calendar-stones, also with beautiful decoration.

It will doubtless be offered as demur against the use of these Aztec designs the fact that they would be costly. They certainly would be when hand-wrought in stone or metal for a pillar-dial, and the pedestal would be costly, too. But simpler materials may be used; the Aztec symbols can be employed for a vertical dial, they can be painted in colors on wood. For the wall of a summer-house or as the decoration on any outbuilding, such as stable, barn, granary, storehouse, I think this Aztec sun-dial would be strikingly ornamental. And any object that has a story, just in that is satisfying.

The earliest-known symbol in the world, the widest-known symbol, and, I think, the most fascinating symbol is the swastika. Extended and varied is its bibliography. The most accurate account of it is the monograph of several hundred pages prepared and printed for the National Museum at Washington. Of this I must tell that it was sent me by an enthusiastic man of science, who wrote, "I believe we have here every existing exponent of the swastika in the known world." I had the pleasure of sending to him, in a few hours after the receipt of his letter, a domestic swastika which was not included in the book: a square of an old patch-work quilt; an everyday design found in old farm-houses in New England, where it is named, in a triumph of irrelevance, Bonaparte's Walk.

Great speculation has been made over the relation between the swastika and the sun, because the

Aztec Calendar-stone.

two signs have been associated by primitive peoples. The sun-symbols of the bronze age were the swastika, the ring-cross, the wheel-cross, indicating the sun-car; the triskele or three-armed cross; the S-shape or sun-snake; in Egypt, the sun-ship.

In the Kensington Museum is a large bronze trumpet found in a bog in Wismar, Germany, near the Baltic Sea. It may have been used in sun-worship, for it is covered with borders and ornaments composed of these sun-symbols.

I would employ as a beautifully symbolic decoration for a sun-dial the sun-signs of this Wismar horn. They are simple; and could easily be stamped with a die, or cut in stone or metal by a very plain workman. It would be a pretty design and a meaning one, and would serve for pedestal and dial-face. The swastika alone would serve as a suitable decoration in all its varying forms. I offer this as a suggestion to some Arts and Crafts Society, that they employ their prentice-hands in making a line of inexpensive sun-dials decorated with the sign of the swastika. It is not the purpose of this book to introduce dial designs; but the lack of originality, or rather the lack of notions of adaptation, in nearly all our dial designs is really surprising. Lack of design! why I can think of half a dozen notions for decoration and design as I sit here writing.

A study of the symbolism and mythology of the aboriginal races of the new world affords ample evidence of the appropriateness of their word-pictures, hieroglyphics, and sculptures as designs for sun-dials. These evidences cannot even be named. The real being of the sun-dial is, of course, based upon the cardinal points, the progress of the sun, light and shade. Among the red men the adoration of the cardinal points was universal; so deep was this adoration, this familiarity, that the Indian ever had the

Seven Ages of Man. From Ancient Block Print.

points of the compass present in his own mind, and upon his tongue as well. His very existence as a hunter depended upon this knowledge. When his slow progress had brought him in the secrets of nature from the motions of the sun to the radicals of arithmetic, he took all knowledge as proofs of the sacredness of the four cardinal points. The world of the Mexicans was also to them a square world, literally hung up by cords at the sides, they thought; the ancient cities of Mexico, Quito, and others were equally quartered; their palaces were all square.

Their temples were built with as due regard for the exact point of the compass as were the churches of old England or the farm-houses of old New England. The government form was quadruplicate. The Inca was lord of the four quarters of the earth. Possession was taken by throwing a stone or a fire-brand to each of the cardinal points. Study any faithful pictures of Indians, and see the ceaseless reiteration of the number four and the cardinal points. In many of the picture-writings even the days of the week are placed north, south, east, and west. The four yearly festivals of the Aztecs and Peruvians, their four points, their invocation of the cardinal points, their mourning for four years, their four ancestors, their four worlds and four ages, — I could multiply these examples. The four gods of the winds were called upon by them as did the prophet Ezekiel call on the four winds in the Bible, as still do the Thibetans, the Chinese, the Parsees, the Brahmins.

The veneration of the cardinal points familiarized

these races with the symbol, which, beyond all others, has fascinated the human mind apart from its religious significance,—the cross. The missionaries of the Church of Rome did not bring the cross to America; they found the races of the new world employing the cross as an ancient emblem. In rain invocations, in fire-making, in mound-building, in many times and places, the cross was used, ever pointing each arm to one of the cardinal points of the compass.

All these things prove to us, as do hundreds of other examples, the dignity of simple numbers and the honored place of mathematics,—of dialling, among other branches; and also show the absolute truth of Kepler's saying that the universe is an harmonious whole, and that numbers, like all nature, are in unison with the mysteries of religion. The beauty of numbers is not revealed to all; favored souls like Frankenstein perceive it everywhere in nature, find the world filled with wonderful and beautiful forms. Had he lived to tell us *how to see*, we, too, might find beauty where now is naught but unmeaning lines. His discovery of the universal principle of pure mathematics, his magic reciprocals and harmonic responses, partake of the charm of magic rather than of mathematics.

The Shakespeare lover will find in the pages of the dramatist infinite variety of suggestion for dial design as for dial motto. None could be more fitting than the "seven ages of man." I should, for a dial, take none of the finished fancies of modern sculptors and painters, but some of the cruder notions of earlier days, such, for instance, as a large block-print in the British Museum (facing page 194). Every line of this is significant and every word.

The verses should be read across the page. They are but doggerel Latin. The stages of man's life have been divided into ten in ancient manuscripts;

Four Seasons' Dial-face.

but Hippocrates (460–357 B.C.) and Proclus (412–485 A.D.) made seven ages. A mosaic on the pavement of the cathedral at Siena, supposed to have been laid in 1476, gives seven ages. This block-print is believed to be older still.

The four seasons offer naturally suggestions for a dial-face. These are known to many nations, and the crude symbolism can be made to fit a dial-face in quaint form. This fine dial-face with the designs of four seasons is made by F. Barker & Son, London.

As four figures supporting a pedestal, the Seasons are seen on the dial at Wroxton Abbey, Oxfordshire.

Sun-dial at Belton House, Lincolnshire; Seat of Earl Brownlow.

The figure of Father Time bearing the dial is significant, and has been often employed on vertical dials, on window-dials, and also cast in lead or bronze, or carved in marble for garden-dials. Here is one of Time and Cupid cut in stone. Real passing of Time has had its effect on the faces, as on human faces, which are now grotesque instead of roguish and severe.

Among the "Naturall" sources of emblematic design for sun-dials, and especially for garden-dials, no more delightful or well-filled stores can be found than in the botanic world. This might be in the direct application of conventionalized ornament from a single plant,—the seed, root, stem, leaf, and flower could all afford detail which would indicate the life of a flower for a year. The Lotus designs offer an already worked out and thoroughly finished scheme of emblematic decoration. From this is an easy step to flower language. A sun-dial might have a floral design which readily could speak in the words of old Dr. Donne in his *Elegie:* —

> "In these the alphabet
> Of flowers; how they devisedly being set
> And bound up, might with speechless secrecy
> Deliver errands mutely and mutually."

Leigh Hunt wrote, with his customary lightness of touch, of

> "Saying all one feels and thinks
> In clever daffodils and pinks;
> In puns of tulips; and in phrases
> Charming for their truth, of daisies
> Uttering as well as silence may,
> The sweetest words the sweetest way."

The maidens of Hindoo and Persian races can easily interpret messages of love indicated by flowers, and other messages can be conveyed with equal exactness. There was a day in France when a springing Violet set at the hour of dawn on a dial's face would read, "Another morn will bring to us Napoleon again!" and floral decoration might be chosen that would speak in full the lesson of the sun-dial as we each interpret it. Messrs. F. Barker & Son of London have an ornate dial-face showing a flower for each month of the year.

An intimate study of the floral art of Japan would afford many suggestions for the floral design on a sun-dial, not only a study of one flower, but its grouping with other flowers and leaves, its manner of growth, its age, its fixed position, and its formal signification, — all would have bearing. Japanesque lines of great simplicity could be used for the pedestal. I can fancy the distinction of such a sun-dial set in an Iris garden or in front of a Wistaria arbor.

Floral Dial-face. F. Barker & Son, London.

In the choice of a plant for supplying decorative motives for a sun-dial in our own country, I have a fancy for native American plants. Many of our common flowers offer fine forms for conventionalized ornament. Let me glance from my window and choose at random from the borders of old-fashioned flowers. Many which we fancy are native came, we discover, from the Orient; but here is one whose name tells a tale of American nativity, while the plant offers also suitable forms for our use. It is the Spiderwort, — *Tradescantia Virginica,* — interesting in our early history as being one of the first of our plants to find a home in England, being carried there from Virginia by the botanical explorer, Tradescant, before 1629. The plant perpetuates his name and that of his father, "an ingenious curious gardener" to Charles I.

The pretty French name is *Ephemerie de Virginie.* It has a peculiar fitness for a sun-dial decoration since the flowers open only in the daytime, and for a single day. The flowering of this plant is of great scientific beauty and perfection; the buds in the umbel hang down, are gracefully recurved; but just before the flower opens in the morning sun they stand erect, and when the flower fades, the withered flower and seeds once more bend down. This should be the motive in the hour markings, and can be exquisitely carried out. Under a magnifying glass the plant has additional beauty in the delicate stamens clothed in silky fibres and the gracefully poised anther; these offer lines for the gnomon. It is such a cheerful, sensible plant in real life; never greedily spreading though perfectly hardy, living happily either in damp or arid ground; it needs no care, but sturdily opens its cheerful blue or white blossoms both in our old gardens and in many a wild location. It has scant medicinal qualities in spite of its name, and a belief that it would cure the bite "of that Great Spider," an imaginary creature of the old herbalists. I was always told as a child that the plant was called Spiderwort because the sharply bent grassy leaves resembled a spider's legs, which they certainly do; they can be modelled in conventional forms for the dial pedestal, to give an effect of great stability.

An ancient writer, Guillim, says of the Columbine, "'Tis pleasing to the eye in respect of its seemly shape." Its seemly shape and the fact that it is a native American plant make it also suitable for the decoration of a sun-dial. Any study of its growth, the form of sepals, petals, stamens, will afford ample suggestions for design.

I find that a true love of flowers is in general allied to a love for the philological derivation of the plant name, and at any rate with a desire to know the history of the plant. Of course I have a special interest in the Tulip because I knew it in Persia when I lived there in my first incarnation in that land of sunlight and flowers.

Many books have been written upon Roses and Lilies. There is a Daffodil book, a Crocus book, but none on the Tulip, though it has had a history worthy of extended record, — a more varied and extended history, indeed, than has the Daffodil. But the Tulip, though admired by all — by nearly all — and certainly beloved of the Dutch people, is not a flower of sentiment, as is the Rose, the Daffodil, the Lilac, the Pansy. The Dutch people showed their admiration for the Tulip in many ways subordinate to the incontrovertible one of placing their liking on a moneyed basis; they used it as the chief of decorative work whether in woodwork, pottery, painting, or embroidery. The

Sun-dial in Yard of Friends' Meeting-house, Germantown, Pennsylvania.

Tulip is a curious design, not only in the elaborate crewel embroidery of bed-hangings and petticoats, but in quilts of patchwork piecing, in homespun and home-woven bed coverlets of linen and wool, and also in a curious knitted stitch used in counterpanes, bed valances, and the like. The Tulip was as omnipresent in worsted and metal within Dutch doors as it flaunted in scarlet and yellow bloom in Dutch borders. It was seen in Dutch metal work, stamped in brass, and wrought in iron; and I have a pewter teapot incised with a Dutch motto and Tulip design.

Among the people of German extraction known commonly as "Pennsylvania Dutch," the Tulip held as honored a place as in Holland itself. On the iron fire-back of the open chimney, on the tiles of the close stove, the Tulip design was ever found. Scant petalled Tulips sprung up around the sturdy four-poster, the family bed; they twined like a vine around these posts defying all rules of botany. They were carved on the wooden bowls and spoons; even the wooden shoes bore a carved Tulip as a tulipette, just as the silken shoe of the English maid bore a silk and lace rosette. Here is a busk carved by a sturdy Pennsylvania Dutch lover; no amorous hearts and darts, no silly love-knots, garland this fierce bar of wood. The Tulip had a better significance for both man and maid. And her linen apron, an apron of strong homespun linen, wears a band in red and blue crewels embroidered; not in Rosebuds or Lilies of the Valley, like this pretty and frail India muslin trumpery of English Cicely; but here, again, is a

Honestone Dial-face from Saxony, with Coats-of-arms. Date, 1760. Owned by Author.

goodly red Tulip with blue and yellow leaves springing from the apron's hem. Truly — to paraphrase an English rhymster, —

> "Her long slit Sleeves, stiff Buske, Puffed Verdingale
> With Tulips thus make her Angelicall."

Tulips of more graceful form were wrought in the clasps of the neck-chains and chatelaines of silver worn by the goodwives of these Pennsylvania settlers, as they rode to the Sunday services of their curious sectaries, or chatted on their neighbors' "stoops" on Saturday night. I have seen a chatelaine key bag, and keys all wrought with a stiff Tulip design. And the clasps of Bible and hymn-book which sometimes hung on these silver chains bore also a design of Cross and Tulip, which was a prime favorite.

Of the "Three R's" of an ordinary education in colonial times, writing was ever the most esteemed, the most imperative; and in general penmanship was fine. To write ill was deemed a disgrace. Spelling was rampantly varied, but writing must be good. It was easy to write with elegance with a quill pen, and whether elegant or inelegant in its results, there was a still greater value: never, so it is asserted, does he who writes with a quill pen have that dread disease of the nerves, writers' cramp. I may add, in passing, another assertion as to writers' cramp: the constant employment of a lead pencil in writing will help sadly to produce that distressing affliction.

Now from all this infinite variety of Tulip design there is certainly ample choice for sun-dial decoration and form, and had I time for the doing of it, I know I could shape out a Tulip sun-dial which would be perfect in a Dutch garden. It should not too closely resemble the Tulips all a-row around it, for that were tiresome; but the severe and scant lines of the dial pedestal should be the long grass lines of the Tulip leaves, and the dial-head should open to show somewhat of a Tulip face.

CHAPTER IX

PEDESTALS AND GNOMONS

"A well-built marble pyramid doth stand
By which spectators know the time o' the day
From beams reflecting of the solar ray;
The basis with ascending steps is graced."

— *The Humours of Covent Garden, 1738.*

"The style is of iron, time is golden.
It passes by like a shadow and returns not."

— *Translation of Italian Sun-dial Motto.*

AMONG materials for the making of dial pedestals, "many are called, but few are chosen." There are many other as suitable materials as wood and common granite. White marble is ever good in a garden when of moderate bulk, and of limit also in the number of pieces shown. As the first qualification is conformed to in a sun-dial, we need only add that white marble is ever good for a dial pedestal if of careful and classic design. The colored marbles of many lands afford a wonderful choice; the least eccentric of these make the finest pedestals, nor should varied marbles be seen in one pedestal, unless of well-studied effect and for special purpose. There is ample color in the garden without adding bizarre effects in colored marble. There are charming Mexican marbles, not the onyx of upholsterers' choice, but red shaded marbles of wonderful veining. Many of our states have individual granites. The exquisite Spanish pink tint of natural alabaster is beautiful; but I fear that stone is too frail for out-door exposure in our climate. An exquisite sun-dial pillar could be cut from the richly tinted cream-colored sandstone of Ottawa, which is used to such effect in the magnificent Houses of Parliament in Ottawa. The beautiful warm red Potsdam and Ohio freestone which forms the adornment of these houses is a glorious stone for a sun-dial; it is an allied tint to the natural alabaster. Those who have seen the perfect towers and buttresses and pinnacles of the Government Houses shining in the true golden light of sunset know that heaven and earth lie very close

Simple Dial in a Worcester Garden.

together at such hours. With a background of greenery, and a choice close setting of dwarf Azalias around the pedestal, chosen with care, in precisely true sandstone and freestone tints a bit *accented*, these would in the blossoming give a color study of great wonder and beauty. And I should like to see this sun-dial in winter glowing in buff and salmon and terra-cotta lines, like a great tinted flower, against its evergreen background.

Terra-cotta Pillar.

The ceramic art offers pedestals for sun-dials. I have seen them of terra-cotta which were satisfactory, though many are too ornate. A very good one, made by Messrs. F. Barker & Son, London, is on this page. They can be made of pottery, both pedestal and dial-face, and several such have been made in our better potteries, with garden seats of corresponding design. Mr. H. R. Mitchell of Haddonfield, New Jersey, makes a very good blue and gray stoneware dial-face. I have also seen Chinese vases and garden seats transformed into dial-pillars, but they have in general rather a make-shift air. A pretty one at Floral Park is shown on one page of this book.

Of course a dial may be set upon a wooden post thrust in the ground, but that forms a pedestal of but shaky position; a sawed off tree-stump is far better; for the dial-face must be exactly horizontal to be of use. A properly set dial is built upon a firm foundation—preferably of brick laid below the frost-line— and the dial-face should be set by spirit-level. The smoothing off of a level face on the upper side of a boulder gives a substantial plane for the dial to be fastened to; and if the boulder is shaped right, it is a very good dial foundation.

> "All clean and bare the stones look now, some light, some dark.
> As year by year goes by, lichens will slowly dot
> And drape them in soft tints; beside them shrubs will grow,—
> The barberry and sweet wild rose; its shiny leaves
> The ivy climbing o'er it will display;
> The clematis its silver floss."

A column laid up of cobble stones in mortar offers a substantial and permanent plane also; and if a few lightly clinging creepers be trained over it, or rarely the closely clinging Japanese Ivy, it can be made very effective. In all these dial pedestals the great striving should be to look and to be firm without being clumsy. One of good effect is shown on page 210. It is in the grounds of Mr. Henry T. Coates' residence at Berwyn, Pennsylvania. Another on a mound of stones is shown on page 211. This is at

Sun-dial and Residence of Henry T. Coates, Esq., Berwyn, Pennsylvania.

the residence of Mr. H. R. Mitchell, at Haddonfield, New Jersey.

I would never, however, when the dial-pillar is fine of design, plant any close-growing creeper that would hide its beauty; above all, the Japanese Ivy, which is the English sparrow of flowers. Where it is made welcome, other creepers are crowded out.

Sometimes an absolutely plain shaft gives great dignity to a sun-dial. Such simple pedestals seem particularly fitting for country churchyards and burial-places; nor are they ill-suited to the old-time flower-garden, when house and fences are of plain lines.

In the Friends' burying-ground, Green and Coulter streets, Germantown, Philadelphia, is a brass dial-face fixed on one such square plain pedestal of gray granite. See facing page 202. The gnomon is set on the

Sun-dial at Haddonfield, New Jersey, in Garden of H. R. Mitchell, Esq.

centre of a metal face engraved with an eight-pointed star, the points indicating the different points of the compass. The outer circle has the hours divided to minutes, and inside that is a table with the corrections for reducing the sun time to mean time for each day of the year. This finely calculated and engraved dial was made by W. & S. Jones, No. 30 Holborn Street, London, but has no date. The assigned date is 1778. On a brass plate fastened to the stone is engraved the noble text: "Our days are as a shadow and there is none abiding" — my favorite motto of all sun-dial legends.

The sun-dial on page 213 has a pretty story. It stands in the garden at Huntercombe Manor, Maidenhead, England, — the garden of E. V. B., the Hon. Mrs. Boyle, author of *A Garden of Pleasure*, *Seven Gardens and a Palace*, and *Sylvana's Letters to an Unknown Friend*. Of all modern writers on the garden now living, Sylvana has to me the truest insight into the spirit of a flower, the purest enjoyment of flower life, the happiest manner of telling of her insight and her enjoyment. Every page of her books is a delight to read and I know that her letters were positively written to me. In one she tells the story of this sun-dial: —

"I send you a likeness of the sun-dial, and here is the story of it, if you care to know. In one sense the sun-dial is old, and in another sense it is quite new. It would do to describe it either way. Years ago at M—— I knew a stone-mason's yard where old stone might be picked up. Here in those days I could often find choice old tombstones and bits of church architecture and old London Bridge parapets, stone balls, etc. My last bargain in the stone-yard was four corners of a tomb of the sort common a hundred or more years ago. For fifteen years I doubted what to do with them, till suddenly one summer day a sun-

Sun-dial at Huntercombe Manor, Maidenhead, England. Residence of Hon. Mrs. E.V. Boyle.

dial was decided. The difficulties of arranging it were great, and the work filled up at least two happy weeks. Three old carved stone tablets with lines in alto relievo were made to grace the sun-dial's head. The lines walk with stately step from the sun rising southwards towards

Sun-dial made from Old Tomb, Huntercombe Manor, Maidenhead, England.

the north. The well-weathered marble brackets on which the gnomon rests had lain in patience under the laurels for many a year till Time brought round an hour and a place for them. And then came the fulfilment of the whole, the motto of a famous architect given to me, and engraved around the upper step, *Lux et umbra vicissim, sed semper amor.* This motto, published in my *Garden of Pleasure*, has been since translated into English and printed in someone else's garden book. I wish you could see the white lily planted two autumns ago just under the lowest step. The slender shaft of its stalk carries now for the first time three or four buds, and when the sun shines upon it a delicate half-transparent shadow slants across the stone. For some reason known to itself this lily grows so small that it is like no mortal lily. Blue Gentian in May sets off the stone edging between it and the turf; and a plant of yellow Clematis flings itself in a light embrace around the central column or pedestal. The flower of it I have never seen. I only know its handsome fluffy seed."

I have given this account of Sylvana's sun-dial at length, not only on account of its absolute charm of diction, but on account of the valuable suggestions it gives for the mounting of sun-dials. What infinite pleasure she has had in comparison to the owner of the costly made-to-order pedestal, in this "home-made" pedestal. Of course it is not home-made either, for the carving is fine and has seen good days ere it came to its better days in the manor garden; but the putting together of the different parts necessitated much thought, and brought infinite gratification, like everything else over which we work long and make a success. I know, were it mine, I should never glance at this sun-dial without a thrill of delight over my handiwork. It is well to use old bits of marble and stonework, or old pillars and pedestals of turned wood, if one can find them of good simple shape. Charles Dickens used a pillar of the balustrade of old Rochester Bridge as a pedestal for his sun-dial at Gadshill. This dial and pillar were recently sold in London for fifty pounds. The early days of the dial are told in an inscription cut upon it. A clergyman of Suffolk, England, has a dial in his vicarage garden at Pakenham set on a part of the balustrade of London Bridge. The bridge was taken down in 1832.

I have known twice of using as a dial pedestal the stone roller of a worn garden or lawn roller of the old-fashioned type. Set on end firmly into the ground, and with a well-designed brass dial-face covering the other end, it was a very satisfactory pillar, and carried with it that pleasant sense of a decorous and not useless end of the days for a faithful old servant, albeit of senseless stone, which one feels also for a worn old mill-stone turned into a doorstep; for a well-curb made into a flower-stand; or for an old Dutch windmill transformed to a house for garden tools.

The Moor, Enfield Old Park, Middlesex.

The richest pedestals are, of course, those of carved figures, suited only to very rich and pretentious gardens. Their cost, whether of marble, stone, metal, or even wood, would prevent their appearance anywhere save in such gardens. A kneeling figure supporting a dial on the head was popular, see page 215. These were sometimes cast in lead. One which stood in the garden of Clement's Inn, and is now in the gardens of the Inner Temple on the Thames embankment, is a negro figure and has been known as "The Moor." It is said that there were in the eighteenth century a number of "statuaries," lead-casters, whose works were between Piccadilly, Park Lane, and Devonshire House; one of these men, John van Nost, made this "Moor" his favorite design.

Sun-dial at Hampton Court.

At Belton House, near Grantham, there is a worn dial in Earl Brownlow's garden, supported by two figures, Old Time and Cupid. This dial, with its quaintly grotesque figures, is shown on page 198. A unique dial is the famous old Turk's Head given on another page. At Windsor, near the Star "Building," stands a sun-dial with a highly carved marble pedestal, which is said to have been the work of that man of infinite genius, Grinling Gibbons. The carving is in high relief, and the Star and Garter is engraved on the dial-face with the motto, *Honi soit qui mal y pense. Henricus Wynne, Londinii, fecit.*

In 1895 a beautiful dial was designed and set up by Messrs. Brewill and Baily at Whatton House near Loughborough. It is here shown. A moulded

Sun-dial at Whatton House, near Loughborough.

circular top is carried by four draped figures of the Muses,—Clio, Euterpe, Erata, and Urania. This group resembles the dial in The Dane John at Canterbury. The exact size of figures for a sun-dial must be guided by the extent of the garden. I do not like colossal figures even in lead, still less do I like to see displayed

> "A little goddikin
> No bigger than a skittle pin,"

as Cotton wrote. On page 414 is shown the dial at Wroxton Abbey, Oxfordshire, described by Mr. Blomfield in his *Formal Garden* as "a moulded circular top carried by four draped female figures, who stand on a square pedestal, the angles of which are decorated with rams' heads and swags of fruits and flowers." Perhaps the camera may be held at fault, but certainly these "Four Seasons" seem dumpy little goddikinesses.

An appropriate pedestal for a substantial dial in a busy town is standing in the enclosure of the gas-works, northeast corner of Twenty-third and Market streets, Philadelphia. It was originally erected by the Market Street Bridge Company, at the western approach of the bridge, as a memorial to those engaged in the construction of the bridge. Later the obelisk was removed to its present position. It is about twelve feet in height, cut clean and true from gray sandstone, and consists of a shaft standing on two steps, supporting a square block of stone, on the four sides of which are cut dials facing the cardinal points of the compass. The whole is surmounted by an urn carved with a burning flame. On the four sides of the obelisk are carved long inscriptions giving a history of the construction, quantities of masonry, etc., used in building the bridge.

Sun-dial and Dial-face at Harlestone House, Northamptonshire.

Facing page 218 is the sun-dial at Harlestone House, Northamptonshire, the residence of the Duchess of Grafton, and a reproduction of a rubbing from the dial-face. It stands on the lawn at the south front of the house, and bears the name and date, Frederic Spencer, August, 1842. Maker, C. W. Dixey, London. Frederic Spencer was the late Earl Spencer, father of the present peer. The pedestal is very simple, but it is a good example of a well-proportioned plain dial-pillar. There are several sun-dials on the grounds at Althorp House, Northamptonshire, the seat of Lord Spencer. One shown on this page has a most elaborately engraved face, telling the time in various parts of the world. Another dial is depicted on page 220. This dial has recently been acquired by Lord Spencer, and placed by him at Althorp House. It formerly stood in the Admiralty House Gardens, Whitehall, London, that memoried spot. When the Admiralty was enlarged last year, the old stone garden house or summer house and this dial had to be removed. Lord Spencer was permitted to purchase both. There was vast appropriateness in their coming into

Sun-dial at Althorp House.

his possession, for he has been First Lord of the Admiralty and his grandfather was also First Lord of the Admiralty as a member of Pitt's administration in those great and glorious days when Lord Nelson ruled the water, at the time of the battle of the Nile and other great naval victories. It is told that Nelson sat often in this garden house with the sun-dial standing before him, talking over naval policy, in those days when England faced down Bonaparte.

Sun-dial from Admiralty Garden.

At Lindfield, Sussex, is Old Place, the seat of Charles E. Kempe, Esq. The house was built originally in 1590 and has been added to wholly in the old spirit. The sun-dial is of an unusual form and decoration, bearing a general resemblance to the dial at Oxford College. The dial-head has been made and set within a few years by Messrs. F. Barker & Son of London. It is a block with four dial-faces raised on a tall pillar, around which twines in large black and gold letters the motto in a spiral reading. The motto runs:—

Tempora prætereunt; nunc sol nunc umbra vicissim
Prætereunt; super est ecce perennis amor.

TIME FLIES, SUNS RISE AND SHADOWS FALL
LET TIME GO BY. LOVE IS FOREVER OVER ALL.

The words *Perennis Amor* are illustrated by a brooding pelican in bronze surmounting all. The

Sun-dial at Yaddo, Saratoga, New York; Country Seat of Spencer Trask, Esq.

pillar itself is raised on a high block covered with ivy, so the whole dial is a very imposing figure.

An entirely different form of support for a dial-face is given on page 221. This beautiful sun-dial is

Bronze Dial-face at Yaddo, Saratoga, New York; Country Seat of Spencer Trask, Esq.

in the Rose garden at Yaddo, near Saratoga, New York, at the country seat of Spencer Trask, Esq. The dial is like an antique table, supported by two carved figures. It is an exact copy of a beautiful

Pillar-dial at Old Place, Lindfield, England; Seat of Charles E. Kempe, Esq.

carving excavated at Pompeii, and it was made for Mr. Trask by express permission of the Italian government. The dial-face is very fine (page 222); it was designed and made by Messrs. F. Barker & Son of London, and bears two exquisite verses by Rev. Dr. Henry Van Dyke, written specially for this dial. One reads:—

HOURS FLY
FLOWERS DIE
NEW DAYS
NEW WAYS
PASS BY:
LOVE STAYS.

The other is at the base of the gnomon:—

TIME IS
TOO SLOW FOR THOSE WHO WAIT
TOO SWIFT FOR THOSE WHO FEAR
TOO LONG FOR THOSE WHO GRIEVE
TOO SHORT FOR THOSE WHO REJOICE
BUT FOR THOSE WHO LOVE
TIME IS
ETERNITY.

The house of Gilbert White at Selborne still stands close to the village highway. Its softly toned bricks and green vineries make it the ideal rural home. The grounds are much the same as during the naturalist's life. In the meadow is his shivering Aspen; and on the green his Sycamore. The brick wall which he built still bears the tablet and date, G. W., 1761. His favorite walk still stretches its narrow brick pathway over The Hanger. Surely, the sun-dial and tortoise must be still here! The lawn is glittering with sunshine so the aged tortoise cannot be seen, but here is the sun-dial on the verge of the lawn, just as he placed it, and read daily its informing letters. You can see its picture here on this page, and a very good model, too, would it form for those who constantly write to me searching for simple well-proportioned dial-pillars. The dial pedestals shown in the illustrations throughout this book offer vast variety of design. Many of them have been chosen and presented simply to instruct the dial seeker.

Sun-dial of Gilbert White, Selborne.

Opposite this page is shown a very satisfactory dial pedestal at the home of Charles F. Jenkins, Esq., in Germantown, Pennsylvania. This pedestal is new; it forms a suitable support for the old dial-face, which belonged originally to Nathaniel Spenser, who lived in Germantown before and during the War of the Revolution. His daughter Hepzibah married, and carried the dial-face to Byberry.

She in turn had a daughter whose married name became Jenkins, and she carried the sun-dial to Gwynedd. Her grandson is the present owner. He rescued the sun-dial of his forebears from a chicken-house with gnomon missing, and after a time that was found. Its inscription, Time waits for No Man, is held to be a punning device on the word gnomon. This dial jest, varied to read, Hours stay for No Man, I wait for No Man, etc., is seen on many English dials.

Two or three years ago a liberal prize was offered in one of our American art associations for the best design for a sun-dial. I know not the specifications in this contest, nor whether there were limitations. I have seen the designs which were deemed the most creditable, and in one case I looked upon the drawing with much curiosity, querying whether the artist had ever seen a sun-dial, or really understood either its significance or its working, as he certainly did not its traditions.

The gnomon of the sun-dial is that piece which projects from the face of the dial, the shadow of which tells the time of the day. It is often triangular, but may be of various shapes; in fact, an obelisk or any index or line which marks a meridian line is also a gnomon. This gnomon is also called a stylus, or style, or stile, or index, or pointer — these all mean precisely the same thing. Florio says, "The gnomon is the gnow-man or know-man of a diall, the shadow whereof pointeth out the howers." From this comes the word *gnomonics*, or as it once was spelled *gnomonicks*, the art or science of dialling; and various other words, such as *gnomonist*, one versed in dialling, and *gnomonology*, a treatise on dialling. The derivative adjective is *gnomonic*, *gnomonical*, and *gnomic*, but as the last-named word has another remote signification, it is not much used. I may say in passing that in the word *dialling* I have clung to the spelling always found in the old treatises and trigonometries; the spelling given in modern dictionaries is with a single l — dialing. On the exact setting of the gnomon all the worth of the sun-dial depends; of course all parts should be exact, but the gnomon *must be* precisely made and set. Therefore it is not well to make the gnomon of wood, because it may warp and twist.

I would suggest to all who are erecting sun-dials, especially horizontal dials in a garden, that more thought and work be spent upon the gnomon than is generally done. Being ordinarily of metal it can be engraved on its flat surface, or, better still, it can be pierced. The use even of a monogram in the design will add to its interest, or a date or crest. I like a large gnomon with as much fine pierced work as can be put upon it. When pierced brass work of such exquisite design was used in old watches, it is strange the brass worker did not turn to the sister timekeeper, the sun-dial, as a field for delicate ornamentation. I have a collection of two hundred old brass verges or bridges from ancient verge-watches in which the designs show every variety of exquisite tracing and outline. I know no gold wrought work to compare with them in delicate beauty, and were they of precious metal, they would make a superb

Sun-dial at Cranford, Germantown, Pennsylvania.
Seat of Charles F. Jenkins, Esq.

necklace. Some such work, though of necessity much heavier and of a deeper cutting, since it is to be exposed to the weather, would I see on the stylus of the sun-dial. It could carry out in finest effect the design of pedestal and face. Or the gnomon might be given a voice and speak both to the dial-face and to that person who is termed in the old dial-mottoes the "Passinger," — that is the passer-by.

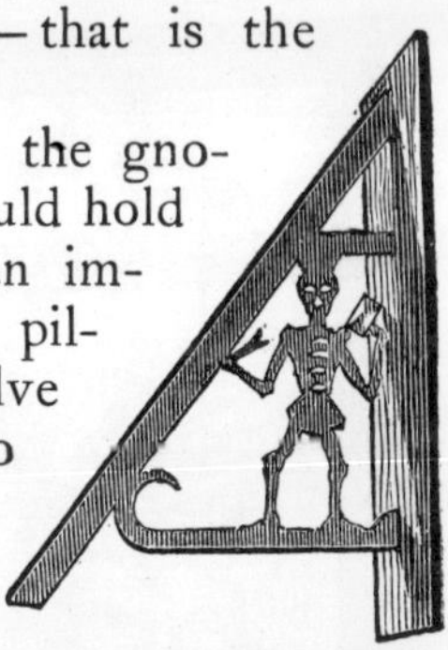

Gnomon of Dial, Lelant Church, Cornwall.

In a facet-headed dial where the gnomons are so prominent, they should hold the chief ornamentation. I can imagine a beautiful dial — a simple pillar supporting a block with twelve faces, each a dial; these faces to have no ornamentation, merely to show the hour lines. The gnomons could be pierced in a floral design, such as the Tulip. Each of the smaller gnomons could be two or three leaves, or a leaf and bud. The four large dials would show the full flower on their gnomons. The pillar should be plain save at the base, where a circular block could show in very low relief a few lines suggestive of Tulip leaves.

Great indifference or lack of taste is often shown in regard to the relation of the ornamentation of the dial-plate to that of the pedestal. The adornment of the plate and sculpture of the pedestal should correspond in design, or, at least, be of a similar school of decoration. You do not wish a Japanesque engraving of lines with a Grecian pedestal;

nor would I wish a floral ornamentation on the dial-plate and the signs of the zodiac on the pedestal.

Very rarely an old gnomon will show some curious design. On page 227 is pictured the gnomon of a vertical dial at Lelant Church, Cornwall; it is the figure of a skeleton standing on a horizontal bar. This is pierced in such a way that his ribs, skull, dart, and hour-glass are plainly seen. This emblem of Death, a skeleton, was held to be as suitable to a sun-dial as to a tombstone; and sometimes the dial bore a carving of skull and bones. One is shown on this page, also page 230. I have seen a mounted globe serving as a gnomon. An elephant's trunk and the wing of a bird have furnished designs for gnomons. A very fine gnomon, shaped like a dragon, is upon a dial made in London for an American garden. It is shown in this book.

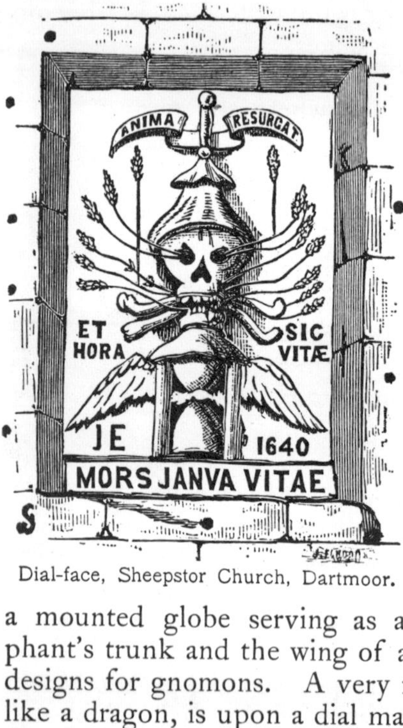

Dial-face, Sheepstor Church, Dartmoor.

A very curious gnomon and a very curious dial was that of the Church of Brou in the Savoy valley.

It is said it was made for the use of the workmen of many lands who built the church.

"Stones are sawing, hammers ringing,
  On the work the bright sun shines,
In the Savoy mountain-meadows,
  By the stream below the pines.

"On her palfry white the Duchess
  Sate, and watched her working train,
Flemish carvers, Lombard gilders,
  German masons, smiths from Spain,"

thus wrote Matthew Arnold in his poem, *The Church of Brou.* This sun-dial was a great circle on the pavement, thirty-three feet in diameter; and the hours were marked in bricks. The time-seeker himself, were he Flemish carver or smith from Spain, formed the gnomon. He placed himself on the spot marked with the name of the current month and his shadow fell on the correct hour. A very elaborate and exact dial was made in Dijon about a hundred years ago by one M. Caumont. Four great blocks of stone marked the points of the compass, and were carved with the signs of the zodiac, and other long slabs of stone with the meridian line and east and west line. Outside these was a circle of twenty-four great stone slabs, each marking an hour. The time-observer set an upright stick on the meridian line opposite the initial letter of the month, and its shadow showed the correct time. I was once shown at a seaside resort a row of numbered stones and a socket, and told to thrust a long pole in the socket, when its shadow would fall on the stones and tell the hour. This we did, and the result proved the pole and stones a very fair timekeeper.

I have a fancy that a sun-dial should ever have some extrinsic value; no object yields more readily to the power of association. Let your dial be made from stone taken from some historic or memorable spot. For instance, a pedestal was cut in stone taken from the field of the battle of Bennington. In that battle took a prominent part a sturdy old farmer from what is now Vermont. His part was prominent — not that he was an officer, but he was a soldier of such exalted enthusiasm and belief in his cause; he was so fearless, so enduring, so bold, though he was seventy years old, that he became a leader in his company through sheer force of his own belief and his expression of it — as many another leader has become. His quaint and fearless sayings are told to this day. He was a blacksmith, and of course with his temperament he was the best blacksmith in the province; and he was proud of his work, as all first-class workmen are. And, what is far rarer, his grandson is proud of it also; and on the fine shaft cut in grand simplicity of shape from this Bennington boulder, he has set as a gnomon a

Sun-dial in Wall of Black Friars' Burial-ground, Perth.

bronze arm wielding a hammer, a splendid piece of work. It fairly speaks to you of his grandfather, the fighting blacksmith, of the certainty of the blows with which he made his way through life, conquering Time because he fearlessly and cheerfully filled it with honest and dignified work.

Pedestals of Dials at Enfield Old House and Chiswick, Middlesex.

Another dial-pillar has a tenderer message: it is laid in cement of sea-worn stones of nearly uniform size and great beauty of tint, which were gathered from the beach, and the very corner of the beach made memorable to the dial-owners as the place where the twain became betrothed; and since the husband is a well-known Shakespearian critic, it is meet that the motto should be a line from Shakespeare's Sonnet CXVI:—

"Love alters not with Time's brief hours and weeks."

A dial-face which I have seen was cast in metal taken from the sheathing of an old battleship, upon which the dial-owner, when a midshipman, had served during our Civil War. Value could come to the dial through its model; it could be shaped like a dial which had been possessed or designed by some one of deserved renown. As an example let me again refer to the sun-dial pictured on page 13, which is a precise reproduction of the sun-dial of Sir Walter Scott. What value this knowledge gives to it! It is a gate opening to us a world of historical and literary memories. One might reproduce the dial of Gilbert White, shown on page 224. It is a bit more ornate than would seem to please Gilbert White's very quiet tastes. He had an interest, we know, in things allied to sun-dials. Read his letter upon building a heliotrope in the garden — two, indeed; one for the summer, the other for the winter solstice. Several who could own a costly dial have reproduced the Queen Mary's dial at Holyrood Palace; others have adopted it in part. I do not, in general, like an alteration of an historical model. The moment it is imitated in part, it has lost its value, — that of exact picturing. Even an association through the selection of a motto is better than no association.

CHAPTER X

THE SETTING OF SUN-DIALS

> "I stand amid ye summer flowers
> To tell ye passing of ye hours;
> When winter steals ye flowers away
> I tell ye passing of their day."
>
> —*Sun-dial Motto*, REV. GREVILLE G. CHESTER, 1860.

HOW readily a sun-dial may be made beautiful—or marred—by its setting! A picturesque or wise setting can do much to atone for or hide an ugly or ill-suited pedestal. Doubtless many of the charming pictures formed by the sun-dial in old gardens came through the judicious and beautifying touch of Time. I am easily influenced by sun-dials. I must acknowledge myself when in each one's presence wholly under its dominion, and dominated a bit, too, by friendship for the owner of the dial. Thus when I am at Twin Beaches, I think a close-set row of English daisies around the circular foundation, a true daisy-wreath, is the sweetest setting any sun-dial could have. At Hillside, in the Shakespeare garden, I know the bunches of the homely old Snow Pinks are wonderfully satisfying, whether tipped with their low-growing sweet-scented pale stars of fragrance or simply standing in clean clumps of grayish green grass-like foliage,—so suited in tint to the color of the dial-stone. At Cranford the pillar of the old Quaker dial is surrounded with Golden Honeysuckle,—for "Time is golden"; and what could be more graceful and appropriate? But when I turn down the Rose walk of the Van Cortlandt manor garden, or walk through the blaze of sunshine and color and perfume of the Burnside garden near my Worcester home, and find the old sun-dials garlanded and surrounded with Roses, then I know that sun-dials and Roses are best of all. So in the garden of Dial House, which is the name of my country home—which is not, but which is ever to be—there my garden-dial, too, will be partly surrounded with Roses. If the sun-dial is set in a Rose garden, it will probably be at the crossing of two paths, whether these be of grass or earth; or it may be on a grass-plot in the centre of the garden. But if Roses are set near it, they should always be low-growing bush Roses, and small of flower; and prettiest of all would be any of the Pompon Roses, the tiny Fairy Rose, or the Pink Burgundy, or the charming Paquerettes; these in their trim little quilled bosses of color and bloom would well adorn the sun-dial.

And in my garden these Rose-bushes must be set with precise regularity around the dial-base,—at the four corners probably; and the Rose-bushes must be

Sun-dial in Lippincott Garden, Germantown, Pennsylvania.

kept of the same size, and a bit formal of shape. I can look at and admire some irregularity of growth and blossom around the sun-dials of others; but in my own garden, and with my own dial, I wish the precision of the laws which rule the dial, and make it the thing it is, to be suggested by some precision of decoration and surroundings.

That the bloom at the dial-base should be small and the encircling vine, if present at all, light in growth is proved by regarding the dial of E. V. B., shown on page 213, where the Clematis is scarce more than an outline, and the Lily a miniature thing — but you cannot doubt their fitness for their place.

If one has garden associations, — and especially childish associations, — and if one has, above all, some tender association of memory with a certain plant, I think it well ever to heed them, and to commemorate them if possible through the planting around a sun-dial. For there, in the presence of that which marks the flight of Time, let the Past be recalled in a permanent manner. It is appropriate to the meaning of a sun-dial.

I always like to see my friends' sun-dials, and this summer I walked down the garden path of a friend to see a pretty sun-dial which had been set on a pedestal made of an ancient granite gate-post. This stood on a square slate base raised a single step; this slate step was carved with initials and dates in the old-time manner of lettering, when the possessions of man and wife — Henry and Alice Earle, for instance — would be marked $\frac{A}{HE}$. This stone step was edged around with Ribbon-grass, with large clumps growing at the four corners. I knew why she had planted it thus, — it was in memory of her childhood in a garden. I looked at my friend in silence, then stooped and gathered two of the blades

Sun-dial with Crimson Rambler in Garden of Mrs. Mabel Osgood Wright, Fairfield, Connecticut.

of green and white grass. And oh, what a wealth of garden memories came to me with the sight and the touch of these grass-blades!

What hours had we spent together as children

striving to find two blades whose colorings and stripes were exactly alike! Nothing hung on the performance of this task; there was no traditional promise of good luck, no dread of uncanny happenings if one failed to accomplish this grass-mating. It was absolutely without reason, yet we hung among the grass-grown bed at the foot of the garden all the long summer afternoon, parting and peering and culling and comparing them. I recalled with delight this garden dalliance, as it came to me with the pleasant touch of the hot green and white gauze ribbons of grass. Gardener's Garters was their quaint old English name. This grass-matching stood on a high plane, on some purely æsthetic principle which no grown folk could fathom, and which I have now forgotten if I ever understood it, but which formed the essence, the spirit, of all childish flower-lore. I wish I could still feel in any accomplishment of mine to-day the gratification which came to me as I seemed to approach success in our childish and meaningless Ribbon-grass play; it was a triumph over all other garden frequenters. Yet I never found two blades that were exactly alike. Throughout the summer and even during the autumnal harvest of golden leaves, which we also tried to match and mate, we turned to the Ribbon-grass. No one can explain the fascination and charm which it held for us.

Another sun-dial has been set in the centre of a circle of Thyme about ten feet in diameter; and around the base of the dial is a row of Golden Thyme in deference to the jest of the old herbalists; and without the Thyme circle is a circular flower bed of sweet-scented herbs, broken only by two openings of paths. This gives an effect not so much of beauty, but of gratification to the sense of smell; whoever walks to the sun-dial has there a vast number of fragrances that he may snip off and bear away. Great bushes of Sweet Briar, Bayberry, Calycanthus, and Southernwood stand on either side of the path entrances; and there are Lemon Verbenas and Fraxinellas and scented Geraniums, including the spicy Nutmeg Geranium beloved of children. Then there are bunches and strips of herbs, not the ranker herbs such as Rue, Sage, Mint, Pennyroyal, Tansy, and Camomile, for these so overwhelm all others, but there is Sweet Basil, and the pretty Burnet, Costmary or Tongue Plant, and a little Lavender, Sweet Cicely, Summer Savory, Woodruff, Tarragon, Rosemary. There were little low hedges of Box around the flower beds, but many declared that the perfume of the Box overwhelmed all other scents, and many did not like it; so to make the sun-dial a beloved resort for all, herbs of universal welcome only were kept. This sun-dial has been planted and set but five years, yet it is astonishing how it has become endeared to visitors as well as the family, partly through the power of associations of scents. One man writes, "I never smell now a bit of Rose Geranium or Verbena without thinking of your sun-dial and sunlight and summer."

A particularly suitable setting for a sun-dial — especially one standing upon a square platform — is to plant Yuccas at the four corners of the dial-base.

Sun-dial in Garden at Avonwood Court, Haverford, Pennsylvania.
Seat of Charles E. Mather, Esq.

Sun-dial in Garden of Henry Souther, Esq., Hartford, Connecticut.

What an effect these Yuccas have thus planted! Their beautiful blooms are those of a miniature Century Plant. They are like a marble statue, so clear and colorless; indeed, they are like ice by night. How fine the starry columns of bloom seen with the sun-dial against a hedge; and when the great flower-stalk is dead and cut away, the cluster of sword-like, spiny leaves is as classic a decoration as the Aloe or Century Plant. I once saw the balustrade of an Italian garden set with a row of Yuccas in full bloom in white marble jars, and their white spires were grand beyond compare. How much more white things tell in the garden than those of other colors, — the white flowered trees, the white blossoming shrubs! White is the high light, the effective point of the garden, just as it is of stained glass; and when white flowers are set near the white marble dial, they all seem a fine study of light and shade. In the daytime the Yucca's column is hung in scentless but graceful bells, and greenish in tint; but now it is night, and the bells open and stand up, full of odor as they are of light. Pale night moths hover round the flowers and float over the dial, lured by the rich fragrance.

"In such a night as this," the vivid moon-night of Shakespeare, — the moon-night of the *Merchant of Venice*, the moon-night of Lorenzo and Jessica, — in such a night as this, when "tipped with silver are all the tree-tops," and all the living scenes of poetry and drama seem near us, — in such a night as this we would be like William Blake, a little mad, and know that there walked with us those whose names we honor, who died centuries ago. Blake had as companions Alexander the Great and Julius Cæsar and Homer; but I would be deeply content in such a night as this to stroll with William Blake himself, and hear him speak of the icelike Yuccas.

I once saw a sun-dial surrounded by a row of

scarlet Tulips in full bloom; it was a cheerful sight for its day and hour, and had a certain fitness in that it was found in the garden of a flower-lover of Dutch descent, who gave due honor of place to the sun-dial by encircling it with a favorite flower. But even in a Dutch garden, the Tulip seems of too fleeting a bloom to seem suitable as a sun-dial setting.

I can well comprehend the longing of a Tulip-lover to place it thus, — as the place of honor in a garden. No distinction was too great to be shown to the Tulip. With an admiration and affection which did not waver for centuries did the Dutch strive to place the Tulip in prominence. I have referred to this at some length in my chapter on Sun-dial Designs.

On page 242 is shown the sun-dial of Horace Howard Furness, Esq., at his home at Wallingford, Pennsylvania. How fine are the long stems and Poppy-seeds which surround this dial! How beautiful must have been these Poppies in bloom! Every minute of the life of a Poppy is beautiful, yet they are seldom a much-loved flower. Yesterday a flower-lover here at East Hampton was asked her favorite flower, and she answered Joan Silverpin, referring to old Gerarde's quaint words about the Poppy, — namely, "Being of many variable colours and of great beautie, although evill smell, our gentlewomen doe call them Jone Silverpin." Constant reference is made by older writers to their vile scent which apparently every one loathed. But nowadays I find many like the smell of a Poppy; I do, and I like to eat the seeds, as I always ate them in childhood. In many countries they are baked with wheaten flour into cakes.

Sun-dial with Poppies in Garden of Horace Howard Furness, Esq., Wallingford, Pennsylvania.

In German bake-shops you may find, and in the so-called "Vienna Bakeries" of our American cities, a certain roll glazed with yolk of egg, and bestrewn plentifully with tiny purplish grains, curious of aspect, but distinctly pleasing, albeit unusual of flavor. This coarse powder or grain is Poppy-seed, and here is a recipe for these cakes: —

"The seeds of white Garden poppies were made into Biskits or Comfits with honey and served up as a Banketting Dish. The rustical peasants of the country were wont to guild or glaze (as it were) the uppermost crust of their loaves of bread with the yolks of eggs, and then to bestrew it with Poppy-seed which would cleave fast to it. They would put them into the oven being thus seasoned which gave a commendable taste to the bread being baked."

You would think that recipe was for our New York rolls, albeit in the wording of a seventeenth century chirurgeon, but it is far older still. It is a translation of a recipe in Pliny's *Natural Historie*; and should you partake freely of these cakes, I doubt not you would feel the opium which must be in the Poppy-seed, and when you slept therefrom, you would dream of ancient Rome.

There is a certain appropriateness in surrounding a sun-dial with flowers which have a subtle air of mystery; they seem suited to the passing of time, to night and day, to all the magic of life. The Poppy has this quality, felt in full by the two greatest students of the very being of the Poppy that the world has known, — Ruskin and Celia Thaxter. The growing Poppies of Celia Thaxter's gardens and the gathered Poppies of her home were miracles of beauty. An arrangement of Poppies on a mantel, which she kept ever fresh throughout the entire summer, seems to have made a lasting picture of glorious beauty upon the minds of all who beheld it.

Many have a distinct indifference or even dislike of planting flowers in the immediate vicinity of a sun-dial; and I am sure it is wholly a modern fashion. You seldom find an ancient garden-dial, if in its original position, with flowers set near it. But when no herbaceous plants are near the dial, shrubs may be planted at a little distance with wonderful effect. I saw recently a sun-dial which stood in a grass-plot in the clear sunlight. In a semicircle, remote enough that they never could shadow the dial-face, were planted shrubs which through careful selection gave to the sun-dial a succession of blossoming companions from early spring till winter found only the scarlet hips of the Japanese Roses. The spring months are readily filled, but there is a period well known to all garden-makers when the sun-dial would have no blossom companions were it not for the Tamarisk, and those faithful relics from old-time gardens, the Althea, or Rose of Sharon.

Of course the very essence and being of the sun-dial lies in ample sunshine, still there may be a certain proximity to trees great and small that will add much to its graceful existence. Three trees of small growth stood near the white marble shaft of one sun-dial; in winter I knew that these small trees were Peach trees, and I knew they would have their day and hour of beauty, but I did not know that they

Sun-dial at Drumthwacket, Princeton, New Jersey; Country Seat of M. Taylor Pyne, Esq.

were double-flowering Peach trees, and that thereby their time of beauty would be so multiplied, quadrupled, in glory. When I saw them in their glorious bloom, they were the first double-flowered Peach blossoms that I had ever seen. There are certain flower-pictures of extraordinary beauty that seem indelibly imprinted on our eyes and brains, — wonderful scenes which we can never forget; this is such a one. I have only to close my eyes on the dullest day in midwinter, on the longest sleepless night, and I see these wonderful irregular mounds of intense pinkness, these masses of flowers with the pure white sun-dial among them. Its warning words of coming night and darkness and death had scant weight in the sight of such beauty which, like all beautiful things, seemed to me, in my first and unreasoning delight, immortal. All this flower-talk opens another line of thought, namely, whether the flowers in the immediate vicinity of the sun-dial should not be carefully regarded as to their relation to the character of the decoration of the sun-dial. But perhaps, in Horatio's words, "'Twere to consider too curiously, to consider so."

Another wonderful background was a row of Pine trees which had been left standing from the old forest when the house was built and the garden planned. Not near enough to shadow basely the dial at midday, but close enough to render useless the markings of the hours of later afternoon, they watched over the dial, and the sound of their branches seemed the very passing voice of Time.

I never hear now the soft musical sighing, the tender low breathing of the Pines without recalling the tree-planting in Hardy's *Woodlanders* — a wonderful description, yet of few words, wherein you smell and hear and see the beautiful young trees the moment they are planted upright. How solemn and weird is that sighing in an old tall forest! It is a distinct third of three notes, formed perhaps by the different height of the trees or by cross-currents

Sun-dial in Garden of the late Hon. William H. Seward, Auburn, New York.

of air. I walked through such a forest last summer, one with grand mast-trees like those marked by the king's broad arrow of old; trees born to be masts and with the tone of the sea in their chords.

And the tree-voices seemed to bear the weight and profundity of the centuries of their lives, — a solemnity that is not sad, but seems filled instead with the essence of a noble life. It is one of the inarticulate nature-sounds that speak more clearly than words.

The voice of the Pines differs at times. Lowell knows the Pine-tree like a brother, knows its moods and its voices: —

> "Pines, if you're blue, are the best friends I know;
> They mope and sigh and share your feelin's so;
> They hush the ground beneath so, too, I swan
> You half forget you've got a body on."

> "Under the yaller pines I house
> When sunshine makes 'em all sweet-scented,
> And hear among their furry boughs
> The baskin' west wind purr contented."

Forests of tall-growing Pines and forests of masts in our harbors alike are disappearing; thus we lose the finest of those beauties given to us simply through the repetition of perpendicular forms. I presume the brother outline, the long ranks of bayonets, will also disappear from our marching armies, and some insignificant little deadly weapon fill the bayonet's place. Yearly are the picturesque elements of our life taken from us. We are given many comforts to replace them, but no work of science or art can ever equal the wondrous natural beauty of the serried Pine trees and Pine masts.

Even this row of a scant dozen Pines guarding the sun-dial has the charm of a succession of upright lines. It is this beauty of perpendicular forms

Sun-dial at Huntercombe Manor, Maidenhead, England; Garden of Hon. Mrs. E. V. Boyle.

that make many plant rows of Hollyhocks as a background to the sun-dial, where they are beautiful as long as they stand in erect lines even though the blossoms are gone and only the "cheeses" remain studding the stalks with their curious forms. There is much dignity in all of the Mallow tribe in the garden, whether they be our beautiful wild Marshmallows, the Hollyhocks, or our friend the Rose of Sharon.

Sun-dial with Peonies at Kenmore, Fredericksburg, Virginia; Home of Betty Washington Lewis.

We are apt to think and speak of a sundial as being suited to a flower garden; but it is equally so for an expanse of lawn, or even to a paved courtyard with no growing flowers. Of course its happiest home is like every one else's in a flower garden. There are certain gardens to which the garden-dial seems wonted and a part thereof; these are specially all old-fashioned gardens, and all formal gardens, and there is a certain type of garden which promises the presence of a sun-dial. It is impossible to formulate a description of such a one, nor can you give any details by which to know of the treasure within. Sometimes the slightest hint will suggest the presence of a sun-dial to you; sometimes you have an inspiration. I was driving along a Long Island road, on the outskirts of a long-settled village, when we passed an old house with grand central chimney flanked by a nebulous growth of greenery of various heights, which suggested evergreen and ancient shrubbery. A hedge stretched across the front of the forecourt, that enclosure which we call the front yard,—a hedge of comparatively old growth for America. An aged negro was trimming this hedge with an old cavalry sword, which he gallantly and skilfully wielded. I know not whether it was the unusual sight of a sword used as hedge shears—and I assure you it proved an excellent one—or the

Sun-dial at Stenton, the Logan House. Presented to the Society of Colonial Dames by Horace J. Smith, Esq., of Germantown, Pennsylvania.

irregular expanse of shrubbery, but I at once suspected the presence within this garden of an old sun-dial; and when we entered, there it was. The wooden pedestal had rotted away, and the poor stump with the rickety dial-face lay prone among the vast Box hedges, hidden save for such undaunted searchers as ourselves. The metal dial-face was fastened by a single rusted screw to the pillar, and twisted about, and was prone face down, with its gnomon thrust in the ground, in an utter abasement and degradation, which resembled in a half-comic manner the grinding of a nose in the dirt; which resemblance, of nose to gnomon, the poet sung in *Cynthia's Revel*, when "her nose was the gnomon of Love's diall, to tell you how runs your heart." I carefully pushed the decaying pedestal from under the edge of the heavy Box and turned the dial-face to the light, and then brushed off the decayed leaves and earth with which it was caked. I read thereon in well-worn letters these ironic words, *Omnibus exemplum et regula*—A PATTERN AND A RULE FOR ALL.

Alas, poor dial! thou wert a pattern and a rule but for a short time and season!

CHAPTER XI

SUN-DIAL MOTTOES

"A Sun-dial motto should be as short as the Posy on a Ring; as Clear as the Sun that shines on the Dial's Face; and as True as Christian Ethics."

ONE thought cannot fail to come to all who read any considerable number of sun-dial mottoes,—a sense of their inherent refinement and grace. They cannot be coarse, nor clumsy, nor scarcely ungracious. Of course they vary in happiness of conception, but all seem refined. I suppose no one would inscribe a motto on a sun-dial until he or she had given ample thought to the wording, and had indeed meditated deeply in order to seize or shape some poetic thought to be a fitting voice for the serious and dignified dial.

I shall not attempt to give a full list of dial mottoes. The curious reader can find them in many languages to the number of sixteen hundred in Mrs. Gatty's *Book of Sun-dials*. Baron Edmund de Rivière published another long list. Early writers on dialling give many suitable mottoes. I had collected sun-dial mottoes in various languages for many years before I saw Mrs. Gatty's book, but I found on comparison that she had nearly all that I had gathered, besides many more; still, I will give here some of the inter-

Sun-dial on Bridge, Philadelphia, Pennsylvania.

esting ones from my collection. Many in the larger gatherings are valueless as a motto for use on ordinary dials.

One almost unvarying characteristic of the sun-dial motto may be noted,— its solemnity. A very few are jocose, a few are cheerful, nearly all are solemn, many are sad, even gloomy. They teach no light lesson of life, but a regard of the passing of every day, every hour, as a serious thing. Biblical texts offer a vast field for culling sun-dial mottoes. The very best to my mind — my favorite motto — is this solemn warning: —

Our days on earth are as a shadow, and there is none abiding. — 1 Chronicles xx. 15.

Opening the Bible wholly at random, after the fashion of the fortune-seekers of old, my eyes fall on these noble lines: —

Truly the light is sweet and a pleasant thing it is for the eyes to behold the sun. — Ecclesiastes xi. 7.

As a servant earnestly desireth the shadow.—Job vii. 2.

All the days of my appointed time will I wait till my change come. — Job xiv. 14.

Behold now is the accepted time. — 2 Corinthians vi. 2.

I have considered the days of old and the years that are past. — Psalm lxxvii. 5.

His time passeth away like a shadow. — Psalm cxliv. 4.

Lord teach us to number our days rightly and to apply our hearts to wisdom. — Psalm xc. 14.

While ye have light, believe in the light. — St. John xxii. 36.

Let there be light; and there was light. — Genesis i. 3.

Man is like a thing of nought. His time passeth away like a shadow. — Psalm cxliv. 4.

Abide with us, O Lord, for it is toward evening. — St. Luke xxiv. 29.

So teach us to number our days that we may apply our hearts unto wisdom. — Psalm xc. 12.

Oh, remember how short my time is. — Psalm lxxxix. 47.

He brought back the shadow by degrees. — 2 Kings xx. 11.

The Lord's name is praised from the rising up of the sun to the going down of the same. — Psalms cxiii. 3.

This last text and parts of it are a favorite choice for mottoes; and in Latin also. In Northampton, England, on the Queen's Cross were four sun-dials, each bearing a few Latin words of this text. This cross is shown on this page in its present condition; it was set up by Edward I in memory of his wife, Eleanor of Castile, and has been sadly tinkered with and the dials removed. It was a frequent motto on French churches. In Kircher's *Ars Magna Lucis et Umbræ* (1646) is a curious exposition of this verse. A great folding plate is given,

Cross at Northampton, England.

having twenty-four dials set in the form of a tree, and four more at each corner. From this tree radiates this verse in thirty-four different languages. On a scroll is the text, *Sicut oliva fructifora in domo Dei.* This plate was intended to have been mounted on a board, and each dial was to have a gnomon affixed, which would then show the time of the day at the places named. The size and shape of each

Pillar-dial in Graveyard at Dean Row, Cheshire, England.

gnomon is carefully given. This would form a splendid triumph in gnomonics.

Some familiar mottoes are seen on many dials. They are certainly common, and some are common-place, but they are suited to their position.

I LABOR HERE WITH ALL MY MIGHT
TO TELL THE HOUR BY DAY AND NIGHT.

Sometimes these lines are added:—

AS CAREFUL, THEN, BE SURE THOU BE,
TO SERVE THY GOD AS I SERVE THEE.

or

IF THOU WILT BE ADVISED BY ME,
I'LL SERVE MY GOD AS I SERVE THEE.

Mrs. Gatty gives an old "clock-paper" which, neatly written and framed in colored paper, was placed under an ancient timepiece:—

HERE MY MRS. BIDS ME STAND
AND MARK THE TIME WITH FAITHFUL HAND;
WHAT IS HER WILL IS MY DELIGHT,
TO TELL THE HOURS BY DAY AND NIGHT.
MRS. BE WISE AND LEARN OF ME
TO SERVE THY GOD AS I SERVE THEE.

A hundred and more years ago the works of a watch were entirely detached from the case, and circular pieces of ornamental paper were placed within the case to protect the works. These discs of paper were known as watch-papers; they were cut in tiny designs, pricked with a pin, painted in water-colors, and inscribed with verses, posies such as were found in posy-rings. One watch-paper which I have has a motto evidently adapted from the motto of the clock-paper:—

IT IS MY WILL AND MY DELIGHT
TO TELL THE HOURS OF DAY AND NIGHT;

and in a manuscript collection of posies for watch-papers is this similar verse:—

HEAR ME TICK AT YOUR COMMAND
AND MARK THE TIME WITH TRUTHFUL HAND
BE THOU WISE AND LEARN OF ME
TO SERVE THY GOD AS I SERVE THEE.

Other everyday mottoes on sun-dials are: —

A CLOCK THE TIME MAY WRONGLY TELL;
I, NEVER, IF THE SUN SHINE WELL.

AS TIME AND HOURS DO PASS AWAY
SO DOTH THE LIFE OF MAN DECAY.

TIME'S GLASS AND SCYTHE
THY LIFE AND DEATH DECLARE.
SPEND WELL THY TIME
AND FOR DEATH PREPARE.

BE THE DAY WEARY,
BE THE DAY LONG,
SOON IT RINGS
TO EVEN SONG.

AS TIME DOTH HASTE,
SO LIFE DOTH WASTE.

LIGHT RULES ME
THE SHADOW, THEE.

Sun-dial at Barncluith, Cadzow Forest, Scotland.

A very numerous "cousinry" of mottoes is that which in Latin runs, *Non numero horas nisi serenas.* This was said to be the favorite dial motto of Tennyson. In its various forms it is doubtless the most popular of all the sun-dial mottoes. In this modification it was chosen by Queen Alexandra for the sun-dial at Sandringham (see page 259), the home for so many years of Edward VII when Prince of Wales: —

LET OTHERS TELL OF STORMS AND SHOWERS,
I'LL ONLY COUNT YOUR SUNNY HOURS.

Vertical Dial at Sandringham, England, residence of King Edward VII of England.

I COUNT THE BRIGHT HOURS ONLY

was on Prince Albert Victor's dial. This was an octagonal pillar with several dials which stood in front of the Exhibition Buildings in Edinburgh in 1886. The Exhibition was opened by the prince and the dial named for him. Other mottoes were: —

AS A SERVANT EARNESTLY DESIRETH THE SHADOW.

LIGHT IS THE SHADOW OF GOD.

TIME AND TIDE TARRY FOR NO MAN.

TIME IS THE CHRYSALIS OF ETERNITY.

WELL ARRANGED TIME IS THE SUREST SIGN OF A WELL ARRANGED MIND.

TIME AS HE PASSES US HAS A DOVE'S WING,
UNSOILED AND SWIFT AND OF A SILKEN SOUND.

TAKE TENT O' TIME ERE TIME BE TENT.

On a fine dial in the Isle of Wight this reads: —

TAK TINT O' TIME ERE TIME TAK TINT O' THEE.

To return to our motto-group. The form on the sun-dial at the fort at Delhi reads: —

I COUNT NONE BUT SUNNY HOURS.

Others are: —

I COUNT THE SUNNY HOURS; I MARK ONLY SUNNY HOURS; I MARK NONE BUT SUNNY HOURS; I NOTE THE BRIGHT HOURS OF DAY; I NUMBER NONE BUT SUNNY HOURS; I ONLY MARK BRIGHT HOURS.

On a slate-dial owned by the Duchess of Cleveland, which was captured by the allied forces in 1854, is this motto: —

I MARK NOT THE HOURS UNLESS THEY BE BRIGHT,
I MARK NOT THE HOURS OF DARKNESS AND NIGHT.
MY PROMISE IS SOLELY TO FOLLOW THE SUN
AND POINT OUT THE COURSE HIS CHARIOT DOTH RUN.

A Latin variant is, *Horas nullus nisi aureas* — I count none but golden hours. This is exquisitely delineated on a vertical dial designed and set by A. G. Humphrey, Esq., at Crowborough Cross, Sussex, on a pole in his garden. The motto and numerals are in open iron work on a semi-transparent gilt ground, which shines out gloriously in the sunlight. Thus the motto has a double meaning.

Sun-dial at Edwin Forrest Home, now Mount St. Vincent Convent.

Another allied motto reads: —

THE HOURS, UNLESS THE HOURS BE BRIGHT,
IT IS NOT MINE TO MARK;
I AM THE PROPHET OF THE LIGHT,
DUMB WHEN THE SUN IS DARK.

In an old album there is written this poem by Professor Samuel Finley Breese Morse, with a note

The Lodge at Charlecote House, Stratford-on-Avon.

saying that he saw the motto on a sun-dial at Worms:—

To A. G. E.

Horas Non Numero Nisi Serenas.

"The sun when it shines on a clear cloudless sky
Marks the time on my disc in figures of light;
If clouds gather o'er me, unheeded they fly,
I note not the hours except they be bright.

"So when I review all the scenes that have past
Between me and thee, be they dark, be they light,

I forget what was dark, the light I hold fast,
I note not the hours except they be bright."

SAMUEL F. B. MORSE.

WASHINGTON, March, 1845.

A sun-dial motto may be simple in its wording and it must be lucid. Lucidity is often confounded with simplicity; but the former is a quality of style, and the latter of thought. A straining after rhyme must not be permitted to make the thought of the motto obscure. For instance, this motto from *Lucile* is pretty, but it is not lucid; in fact, it is not true:—

THE DIAL
RECEIVES MANY SHADES, AND EACH POINTS TO THE SUN,
THE SHADOWS ARE MANY, THE SUNLIGHT IS ONE.

On the sun-dial of Thornby Church, Northamptonshire, are these serious lines:—

MARK WELL MY SHADE, AND SERIOUSLY ATTEND
THE COMMON LESSON OF A SILENT FRIEND,
FOR TIME AND LIFE SPEED RAPIDLY AWAY;
NEITHER CAN YOU RECALL THE FORMER DAY.
YOU ARE NOT ABLE TO RECALL THE PAST,
BUT LIVE THOU THIS DAY AS IF THE LAST.

At Oxford there is a sun-dial bearing the arms of Thomas, Earl of Wharton, who was Lord-lieutenant of Oxfordshire from 1691 to 1702; it bears these clever lines:—

A MOMENT—MARK HOW SMALL A SPACE
THE DIAL SHOWS UPON THE FACE;
YET WASTE BUT ONE—AND YOU WILL SEE
OF HOW GREAT MOMENT IT CAN BE.

One of the most exquisite and perfect of all antique English sun-dials partakes of the lectern-shaped type. It is shown facing this page. This dial is at Moccas Court, Herefordshire, the seat of Rev. Sir George Cornewall, Bart. It has many fine mottoes;

Sun-dial at Brockenhurst Park, Hants.

in Latin is a verse of the Nineteenth Psalm, "The heavens declare the glory of God, and the firmament showeth his handiwork." Also in Latin is a text from Deuteronomy sixth, "Love the Lord thy God with all thy heart." Other mottoes are: *Instar globi stat machina mundi* — Like a ball stands the frame-

Sun-dial at Moccas Court, Herefordshire, England.
Seat of Rev. Sir George Cornewall, Bart.

work of the world; *Si culpare velis, culpabilis esse cavebis. Nemo sine crimine vivit: idcirco ne temere judicto*—If thou wouldst blame, thou wilt beware of being blameworthy. No one lives without reproach, therefore judge not rashly; *Sol est lux et gloria mundi*—The sun is the light and glory of the world.

There is also this quaint English verse:—

TYME PASSETH AND SPEAKETH NOT,
DETH COMETH AND WARNETH NOT,
AMENDE TODAY AND SLACK NOT,
TOMORROW THYSELF CANNOT.

This dial is thought to be of the time of Charles II.

Mr. Evans has a very interesting Manx dial made of marble, which has several fine inscriptions. One is most quaint and old-fashioned, and is a favorite of mine:—

WHILST PHŒBUS ON ME SHINES,
THEN VIEW MY SHADE AND LINES.

There is a group of Latin mottoes which are often found: *Sic transit hora*—Thus passes the hour; *Sic tempora labuntur*—Thus glides time; *Sic transit gloria mundi*—Thus passes the glory of the world; *Sic transimus omnes*—Thus pass we all; *Sic transit hora*—Thus passes the hour; *Sic vita*—So is life; *Sic vita fugit*—Thus life flies; *Sic vita transit*—So life passes. With many variants these are seen on many English churches and houses, and on garden-dials in Scotland, England, France, Holland, and Italy. *Sic transit glori mundi* is upon a brass sun-dial at Matale, Ceylon, engraved in curious characters, the initials being shaped like animals.

On the tower of Shillington Church, Bedfordshire, a clock and sun-dial were formerly found with two exceptionally happy mottoes. The sun-dial had this: *Sine soleo sileo;* and the clock: *Sine sole loquor.*

A sun-dial may speak in solemn voice and yet not be offensively despairing. I particularly dislike such mottoes as this on the dial at Brougham Hall, Westmoreland:—

O WRETCHED MAN REMEMBER THOU MUST DIE,
SENCE ALL THINGS PASSE AND NOTHINGE CERTAIN BE.

The date cut on this dial is 1660, and at that time and in that condition of English history there were many to whom thoughts of death and solemn warnings and dread of hell were as the breath of life. A skull and hour-glass further decorate this dial. Fortunately it bears on another face the beautiful and appropriate words, *Tempus ut umbra præterit*—Time passes by as a shadow; and also that dignified but most common of all dial mottoes, *Ut hora sic vita*—Life is as an hour.

I own a handsome brass sun-dial about a foot in diameter which bears the date 1748, and these lines, evidently added at a later date:—

HASTE, TRAVELLER, ON THY WAY,
THE SUN IS SINKING LOW.
HE SHALL RETURN AGAIN,
BUT NEVER THOU.

This always seemed to me an ungracious and inhospitable answer to the chance passer-by, who sociably halted to learn the time o' the day; but I find a similar sentiment conveyed in many dial mottoes, — a request not to dawdle around, — and likewise a solemn warning to lose no time thus, since the return of the sun might not bring back the day to the dial reader as to the dial.

Vertical Sun-dial at Germantown, Pennsylvania.

E. V. B. in her book *A Garden of Pleasure* tells of a beautiful motto of allied thought upon a dial in a Riviera garden: —

> "*Io vade e vengo ogni giorno*
> *Ma tu audrai senza ritorno.*"

> "I go and come every day,
> But thou shalt go without returning."

A particularly fine motto is this: *Transit umbra; lux permanet*—The shadow passes; light remains.

Sun-dial at Canon's Ashby, Northamptonshire; Seat of Sir Henry Dryden, Bart.

Its simplicity increases its force, and the sentiment is grateful to every one.

WHEN THOU DOST LOOK UPON MY FACE,
  TO LEARN THE TIME OF DAY,
THINK HOW MY SHADOW KEEPS ITS PACE,
  AS THY LIFE FLIES AWAY.
TAKE, MORTAL, THIS ADVICE FROM ME,
  AND SO RESOLVE TO SPEND
THY LIFE ON EARTH, THAT HEAVEN SHALL BE
  THY HOME, WHEN TIME SHALL END.

This was taken from the sun-dial on or near Dromore Castle, County Kerry, Ireland; its date is 1871.

A severe motto reads:—

I NOTE THE TIME THAT YOU WASTE.

A very spirited motto is in Latin: *Horam sole nolente nego*—I tell not the hour when the sun will not.

In the Ulster *Journal of Archæology* for October, 1901, is a fine description of an interesting old sun-dial from the parish church at Bangor. It is of slate, elaborately carved on both sides, and was set in a sloping position with the outside circular edge tending upward. It has several curious inscriptions, one being the old almanac rhyme,—

THIRTIE DAYES HATH SEPTEMBER,
APRIL, JUNE, AND NOVEMBER;
FEBRUARIE HATH 28 ALONE,
AND ALL THE REST 30 AND ONE.
1630. DEC.

On page 250 is an old sun-dial which now stands in the garden of the Logan Mansion, Stenton, the house now occupied by the Society of Colonial Dames. The dial was given to the Society by Horace J. Smith, Esq., of Germantown, one of the few descendants of the Logan family. On one side are the incised words, WE MUST —(*scil* dial, *i.e.*, die-all). This clumsy joke is common on English dials. It appears under a mural sun-dial in the engraving of Hogarth's picture of *Chairing the Member*. The painting had the fine motto, *Pulvis et umbra sumus* — We are dust and shadows; well suited to the skull and cross-bones accompanying it. But the engraver evidently thought himself a better humorist than the painter, and replaced the Latin motto with WE MUST —. The joke is older than Hogarth. The *Horologiographia Optica* by one Morgan, published in 1652, ends with it. On a church dial is this verse: —

LIFE'S BUT A SHADOW,
MAN'S BUT DUST;
THIS DIALL SAYS
DY ALL WE MUST.

There is a very quaint variant of this motto on a farm-house dial at Millrigy, near Penroth, in the form of a dialogue between the Sun-dial and the Passer-by: —

Diall. STAIE PASSINGER.
TELL ME MY NAME,
THY NATURE.

Pass. THY NAME IS DIE
ALL. I AM A MORTALL
CREATURE.

Sun-dial at Ivy Lodge, Germantown, Pennsylvania.
Seat of Horace J. Smith, Esq.

Diall. SINCE MY NAME
AND THY NATURE
SOE AGREE,
THINK ON THYSELF
WHEN THOU LOOKEST
UPON ME.

Another beautiful dial, with musing figure pointing to the dial-face, is at Ivy Lodge, the home of Horace J. Smith, Esq., Germantown, Pennsylvania. It is shown facing page 270.

On French dials there is a jocose motto which is not uncommon. A cock is painted on the dial-face and the words, *Je chanterai quand tu sonneras*; or, *Lorsque tu sonneras je chante.* This is, of course, the challenge of the silent cock to the silent dial.

At Linburn, Midlothian, Scotland, Ebenezer Erskine Scott, Esq., erected two very fine modern dials. One, shown facing page 172, is an obelisk-shaped dial of good proportions. The other, facing page 274, is a facet-headed dial of great beauty. Both are set on octagonal raised steps. The latter dial is nine feet in height, and was designed by Thomas Ross, Esq., F.S.A. On the upper step is engraved verse 3 of Psalm cxiii., "From the rising of the sun to the going down of the same, the Lord's name is to be praised." On the lower steps are graceful verses which have a curious story; they run thus: —

AMIDDST YE FLOWRES
I TELL YE HOWRES.

TIME WANES AWAYE
AS FLOWRES DECAYE.

BEYOND YE TOMBE
FFRESHE FFLOWERETS BLOOME.

SOE MAN SHALL RYSE
ABOVE YE SKYES.

These mottoes were written by Rev. Greville J. Chester, and appear in his novel *Aurelia*, in his description of the bishop's garden, with "a double

Sun-dial in Mrs, Bell's Garden, Cheshire, England.

row of Hollyhock, spires of flame and rose-color, and white and crimson; and bunches of Golden Aaron's Rod, and Canterbury Bells, and Bee Larkspur, and Prince's Feathers; and later on in the year tufts of purple golden-eyed Michaelmas Daisies: and at the end of all, upon a lump of turf, stood a gray time-tinged sun-dial, inscribed on its four sides with the quaint distiches devised by Bishop Edmund Redyngton, who set it up A.D. 1665."

So vivid was this description that many readers placed implicit confidence in the reality of the old sun-dial and its ancient verses, and the lines have been copied on others than the Linburn dial.

There are two old sun-dials in California. One is in the ancient Mission of San Juan Bautista, San Benito County; it was brought by pious padres from Spain in 1794, and is the official clock of the Mission. The other was set up at Mare Island in 1854 by Admiral Farragut. Its motto runs, *Como la sombra huye la hora* — Like the shadow flies the hour.

Many English poets have had the writing of dial mottoes, and many verses of English poetry have served as mottoes. Dr. Watts wrote a characteristically gloomy verse for Lady Almy at Newington in 1735: —

SO ROLLS THE SUN, SO WEARS THE DAY
AND MEASURES OUT LIFE'S PAINFUL WAY;
THROUGH SHIFTING SCENES OF SHADE AND LIGHT
TO ENDLESS DAY OR ENDLESS NIGHT.

A favorite verse of mine was written by Walter Savage Landor, but I do not know whether it has ever been used on a dial.

IN HIS OWN IMAGE THE CREATOR MADE
HIS OWN PURE SUNBEAM QUICKENED THEE — O MAN!
THOU BREATHING DIAL! SINCE THY NAME BEGAN
THE PRESENT HOUR WAS EVER MARKED BY SHADE.

I wonder whether Chaucer's lines have been set on a dial: —

"For tho' we sleep, or wake, or rome, or ride,
Ay fleeth the time, it will no man abide."

or Spenser's fine line: —

"None can call again the passed time."

Another line which suggests itself as appropriate for a sun-dial is Tennyson's line in *The Ancient Sage*: —

"Make the passing shadow serve thy will."

How exquisite are the lines of the "Prince of Poets," Ronsard: —

"*Le temps s'en va, le temps s'en va, madame!*
Las! le temps non: mais nous nous en allons."

Austin Dobson thus renders them: —

"Time goes, you say? Ah, no!
Alas, Time stays, *we* go!"

Hudibras furnishes this couplet for several English dials: —

Facet-headed Garden-dial at Linburn, Midlothian, Scotland.

AS TRUE AS THE DIAL TO THE SUN
ALTHOUGH IT BE NOT SHONE UPON.

And Addison's *Paraphrase of the Nineteenth Psalm* gives these two lines: —

THOU ART, O LORD, THE
LIFE AND LIGHT
OF ALL THIS WONDROUS
WORLD WE SEE.

From Shakespeare's seventy-seventh Sonnet are these lines: —

THOU BY THE DIAL'S
SHADY STEALTH MAY
KNOW
TIME'S THIEVISH PROG-
RESS TO ETERNITY.

Sun-dial at Bramhall, Cheshire.

Quarles's *Emblems* furnish several whining verses for mottoes. Emblem Number 13, Book III, is a sun-dial. The Bible verse is from Job: "Are not my days few? Cease then and let me alone, that I may bewail myself a little." These lines are his verses and are printed on an English dial: —

READ ON THIS DIAL HOW THE SHADES DEVOUR
MY SHORT-LIVED WINTER'S DAY. HOUR EATS UP HOUR
ALAS! THE TOTAL'S BUT FROM EIGHT TO FOUR.

Another stanza has also been used: —

NOR DO I BEG THIS SLENDER NICHE, TO WHILE
MY TIME AWAY, OR SAFELY TO BEGUILE
MY THOUGHTS WITH JOY — THERE'S NOTHING WORTH A SMILE.

Dial in Rose Garden at Broughton Castle.

TIME FLIES. LINES RISE AND SHADOWS FALL
LET IT PASS BY
LOVE REIGNS FOREVER OVER ALL.

These lines are on a sun-dial owned by Lord Ronald Gower; they are the English rendering of the Latin verses which are on the dial at Old Place, Lindfield, Sussex, see page facing 226.

Far more beautiful are the lines by Dr. Henry Van Dyke given on page 223.

Mr. Evans gives me a group of sun-dial mottoes which are not published in Mrs. Gatty's book, nor in its latest edition by Mrs. Eden. I cannot give them all in full. *Amour pour Amour*, on an ivory portable quadrant dial, French, in Mr. Evans's collection.

AS THE SUN RUNS
SO DEATH COMES,

on a horizontal dial made by "Adam Stear, 1660," belonging now to Rev. G. W. W. Minnes, The Cliff, Weston, Southampton. *Curriculum meum perficiam donec advenerint dominus* — I WILL RUN MY COURSE UNTIL THE LORD SHALL COME, on a German dial of gilt-brass resembling an astrolabe, sixteenth century. *Justum et æquum* — JUST AND FAIR, 1717, and *lucet omnibus* — IT SHINES FULLY, on a large French portable dial in Mr. Evans's collection. *Non sibi soli vivere sed et aliis proficire vult* — IT WISHES TO LIVE NOT FOR ITSELF ALONE, BUT ALSO TO BE OF USE TO OTHERS, on a brass portable dial, probably Spanish, by F. Hieronimus de Arresse, 1598.

And on a German portable dial of brass, 1612, with a nocturnal dial at the back, are these German legends: —

> "*Der Zeiger sol gerucket sein*
> *Wol auf der sonnen grad allein*
> *Das instrument solt hangen lan*
> *Zeigt der schat dir die dages stund an.*"

On the back are these lines: —

> "*Dis schiblein mit dem lengsten zan*
> *Sol auf den tag des monats stan*
> *Durdis mittel loch den potu sihe*
> *Die regel auf den 'Kochal' zihe*
> *Dan wirdt zu handt*
> *Die nacht stunt behant.*"

When translated these run: —

> "The pointer is to be adjusted
> To the degree of the sun alone;

(When) the instrument is hung
The shadow will indicate to thee the hour of the day."

"The disc with the longest notch
Should mark the day of the month,
Observe the pole through the central opening,
Draw the movable rod to the peep-hole,
Then the hour of the night
Will be immediately revealed to you."

CHAPTER XII

THE SUN-DIAL AS A MEMORIAL

"Even the dial, that stood on a hillock among the departed
(There full a hundred years had it stood) was embellished with
blossoms
Like to the patriarch hoary, the sage of his kith and his hamlet,
Who on his birthday is crowned by children and children's children.
So stood the ancient prophet, and mute with his pencil of iron
Marked on the tablet of stone, and measured the time and its
changes,
While around at his feet an eternity slumbered in quiet."

— *Children of the Lord's Supper.* *Translation by* LONGFELLOW.

A FRIEND once said to me that she did not like a sun-dial in a garden nor any statue, or even a seat of white marble, because it seemed to suggest to her a cemetery. There is to me nothing sombre in a sun-dial, and yet it is certainly a meet furnishing of a burying-ground. We all know that Howard the philanthropist said on his death-bed that he wished a sun-dial over his grave. Sir William Temple ordered that his heart should be placed in a silver box and buried under the sun-dial in his garden at Moor Park. This was for love of his garden. I know I could not be buried in a garden though it would be pleasant to sleep there, but I should, I am sure, like a sun-dial over my grave. In Ireland the old graveyards have many ancient sun-dials, in fact nearly all the old dials in Ireland are on tombstones. In many of these the gnomon is missing; and the empty hole in which the gnomon once was fastened is said to have been used as a betrothal hole. The visits of a priest were rare in remote communities, and when he was not at hand to solemnize any marriage, bride and groom together placed a finger in the empty hole in the dial-face and pledged marriage in the presence of witnesses. Whenever the priest appeared he always remarried them in good, legal, and religious form. This empty hole was also called a swearing-hole.

Cross-dial in West Laurel Cemetery, Philadelphia.

There is shown on this page an old sun-dial in the shape of a cross which stands in West Laurel

Cemetery, Philadelphia. It is in a neglected spot, much grown up with shrubbery, and is seldom noted by passers-by. This form of dial is particularly suited to a churchyard or a cemetery; for the form of a cross should not be used lightly. I would not care to see a cross-dial in a garden, though several English clergymen have them.

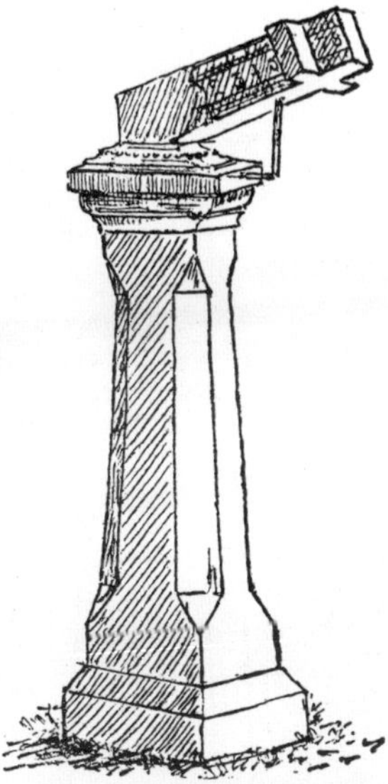

Cross-dial at Scotscraig, Scotland.

Rev. R. W. Essington has erected several cross-dials: one at Shenstone vicarage, another in the churchyard, another at Newquay, Cornwall; and he wrote these lines as a motto: —

> IF O'ER THE DIAL GLIDES A SHADE, REDEEM
> THE TIME, FOR LO, IT PASSES LIKE A DREAM;
> BUT IF 'TIS ALL A BLANK THEN MARK THE LOSS
> OF HOURS UNBLESSED BY SHADOWS FROM THE CROSS.

On another dial he inscribed: —

> THE HOURS ARE GRAVEN ROUND THE CROSSES SIDES,
> AND ON THEM ALL IN TURN THE SHADOW GLIDES;
> IF THE SUN SHINES AND DRAWS A LINE, REDEEM
> THE TIME, FOR LO, IT PASSES LIKE A DREAM;
> BUT IF THE LINE BE ABSENT MARK THE LOSS
> OF HOURS NOT RULED BY SHADOWS FROM THE CROSS.

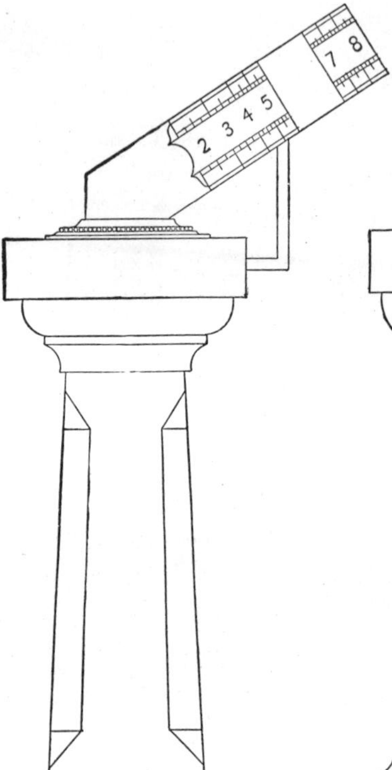

WEST SIDE SHOWING DIAL
AND PEDESTAL AS THEY EXIST

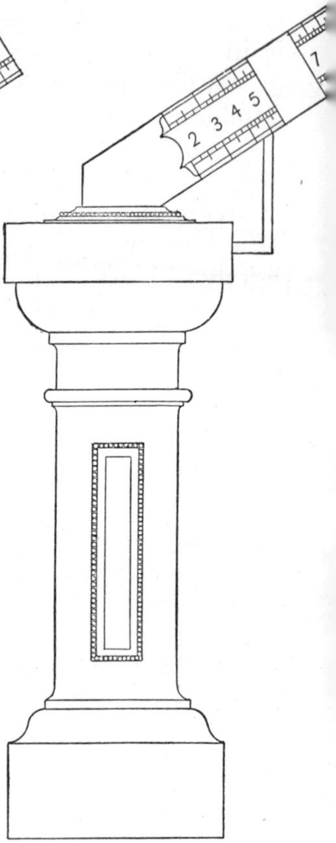

SUGGESTED PEDESTAL AS BEING
MORE IN KEEPING WITH THE DIAL
THAN THE PRESENT PEDESTAL

On a slate step at the base of the dial in the vicarage garden are two mottoes, one in Greek, one in Hebrew, meaning, "The cross gives the hour in sunshine," and "Let there be light." On the shaft are the Latin lines:—

> "*Solis adit lux*
> *Hic docet umbræ crux*
> *Datur hora*
> *Umbram addit nox*
> *Hinc abit umbræ vox*
> *Abit hora absit mora.*"

Another motto reads:—

O COUNT THE HOURS AS ONE BY ONE THEY FLY
AND STAMP THE CROSS UPON THEM ERE THEY DIE.

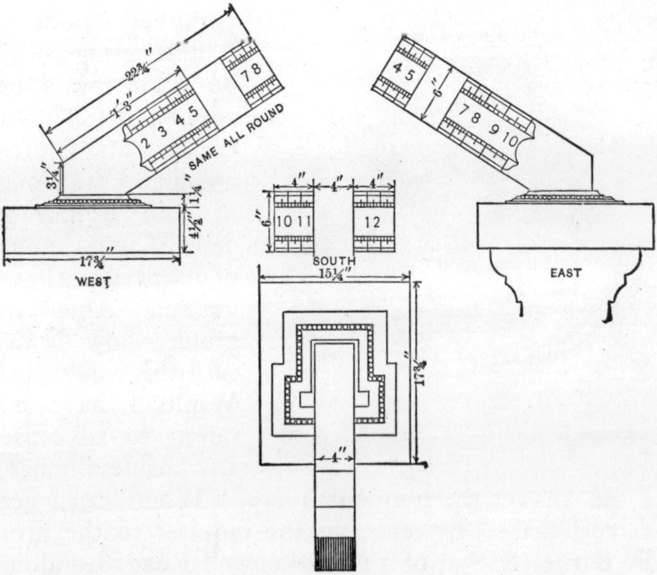

Scotscraig Cross-dial.

Hoping that the thought of a cross-dial will appeal to many who have in their hearts the intent of some memorial, I give on pages 282, 283, as aid in making such a dial, drawings of the ancient cross-dial at Scotscraig, which is of exceptional proportions and beauty. It is supposed to have been made in the seventeenth century. These drawings were generously made for me by Mr. T. S. Robertson, architect, of Dundee, Scotland. And I am sure my thanks to him will be echoed by my readers. Every elevation is given, and in addition a proposed pedestal, in perfect keeping with the design of the moulding at the top. The present pedestal of this cross-dial is comparatively modern. The lettering could be upon brass or copper plates, but the lines and figures are cut in the stone of the Scotscraig dial. Cut in some of our perfect, close-grained American granite how beautiful this would be! While I have no intent to advertise any business house, I can answer the frequent query, "Where can I get a cross-dial?" by referring the inquirer to the firm F. Barker & Son of 12 Clerkenwell Road, London. Their shop at the above address is but a three min-

Brass Cross-dial. F. Barker & Son, London.

utes' walk north of the Aldersgate Street Railway Station, and there can be seen many dial-faces which are not in their catalogue, and occasionally they can supply old dials as well. A beautiful universal portable cross-dial made for this firm for a traveller in India is shown on page 284.

At Newhall, Penecuik, in the Midlothian, stands a sun-dial which is a memorial to Allan Ramsay. On the shaft are five panels bearing carved designs relating to Ramsay's *Gentle Shepherd*. The sixth panel is inscribed: "Here Alexander Penecuik of Newhall, M.D., is said to have given Allan Ramsay the Plot of His Celebrated Pastoral Comedy of the 'Gentle Shepherd.'" On the eighth panel are the words: "Here Allan Ramsay recited to his Distinguished and Literary Patrons, as he Proceeded with them, the Scenes of his Unequalled Pastoral Comedy, amid the Objects and the Characters introduced into it." The eighth panel has a motto:—

OBSERVE HOW FAST TIME HURRIES PAST
THEN USE EACH HOUR WHILE IN YOUR POWER
FOR COVER THE SUN, BUT TIME FLIES ON
PROCEEDING EVER, RETURNING NEVER.
R. B. 1810.

Memorials need not be always of the dead, but of the living. They may indicate some hero-worship, some literature-love. I was much impressed at the home of Mrs. Mabel Osgood Wright with the indelible impress, not of his life alone, but of his character, which her father had left upon his country home. A stranger could walk around the grounds and woods and know the man who loved them; know, too, the great names he honored, — for here the names are chiselled on a rough boulder in the

Pillar-dial, Appleby.

grove; know his love of Dante, that in this sequestered spot he gathered every year on Dante's birthday even the little children, while he read to them passages from the pages he loved. We should have more Dante lovers had we such leaders and such inscriptions of verses from the poet. Here on this vast stone pulpit are dates, for here he preached, during the stirring days of the Civil War, patriotic

The Countess's Pillar.

sermons to audiences which no church walls in the village were broad enough to hold. Here might be inscribed the victories for which he thanked God.

On page 286 is shown one of the most interesting historical sun-dials in England. The dial stands on a tall pillar at the top of the hill, in the village of Appleby, and is one of two erected by the famous Anne Clifford, Countess of Pembroke, Dorset, and Montgomery, who lived from 1590 to 1675. She was heiress of several involved estates which she triumphantly bore out of their tangles into prosperity, fighting throughout the courts with a zest, and ruling these estates with equal vigor and great discretion. Among other victories she claimed and held throughout her life the office of hereditary high sheriff of Westmoreland, in spite of the opposition of King James. She had disputes with the crown under King Charles, and defended her castles against Cromwell. She repaired her crumbling castles, restored and built churches, and erected other monuments — and "never tasted wine or physic" — a fine anti-climax. Within the Appleby Church is her monument, showing her life-sized, sturdy figure.

On this dial is the motto, "Retain your loyalty, preserve your rights," which, in the light thrown by the events of her life, showed that she chose a motto which was to her of deep feeling. It is also significant when taken in connection with an anecdote of her told by Horace Walpole. Its authenticity has been questioned, but it is wonderfully characteristic. She had been communicated with by an officer of the king, and told to support a certain candidate for high office in her borough. Her answer ran thus: —

"I have been bullied by an Usurper, I have been neglected by a Court, but I won't be dictated to by a Subject. Your Man sha'n't stand.

"ANN DORSET PEMBROKE AND MONTGOMERY."

On page 287 is shown "The Countess's Pillar,"—a pillar set up also by her on the wayside between Brougham and Appleby. The square block which surmounts the octagonal pillar has carvings on four sides; two are sun-dials. It is here given in enlarged size to show the detail. There is also an inscription which explains the name given the structure:—

Dial-block of the Countess's Pillar.

"This Pillar was Erected Anno Domini 1656 by the Right Hon. Ann Countess Dowager of Pembroke and Sole Heir of the Right Honorable George Earl of Cumberland, etc., for a Memorial of Her Last Parting in This Place with Her Good and Pious Mother The Right Honorable Margaret Countess Dowager of Cumberland, the Second of April, 1616. In Memory

Whereof She also left an Annuity of Four Pounds to be Distributed to the Poor within the Parish of Brougham Every Second Day of April For Ever upon this Stone Table. Laus Deo."

A custom existed in the sixteenth, seventeenth, and indeed the eighteenth centuries, of accompanying a departing guest some distance on his way—going "agatewards," it was called. And it was deemed a great indignity if no one rode out with the traveller. Sometimes a great party would go on horseback several miles on the way. Doubtless the countess parted with her mother by the wayside, after some visit which one had made to the other.

The stone table for the alms still stands, but is concealed in the picture by the ugly protecting fence. The present rector of Appleby Church was asked whether the alms was still given, and answered, "I'd hear from it if it wasn't."

Another tender memorial to a loved mother is the sun-dial on page 291. It is at Neaum Crag, Ambleside, and was erected by Albert Fleming, Esq. The motto is the Latin rendering of the text, Psalm cxxxix. 12, "The darkness and light to thee are both alike." On the base of the pedestal are the noble lines of Matthew Arnold:—

> "O strong soul, by what shore
> Tarriest thou now? For that force,
> Surely, has not been left vain!
> Somewhere, surely, afar,
> In the sounding labour-house vast
> Of being, is practised that strength,
> Zealous, beneficent, firm."

Sun-dial on Library Tower, Princeton University, Princeton, New Jersey.

I often wonder when I see the very trying results of affectionate desire to erect a memorial to our beloved dead that one dignified form is wholly neglected, especially when such unhappy examples

Dial with Dipleidoscope, Neaum Grag, Ambleside, England.

of stained glass are set in our churches — I mean a memorial window-dial of stained glass. Window-dials are described in Chapter II. A suggestive and beautiful design would be of some figure, an angel or some Biblical character, holding a pointer to the hours. *Dante's Amor*, page 16, the angel on the Chartres Cathedral, page 18, and the angel with dial upon the Genoa Cathedral, facing page 14, would all be exquisite designs for window-dials. Of course

Sun-dial at Yale University, New Haven, Connecticut.

such a dial would be read from the inside of the building, and thus the hours would be numbered, as on a clock, from right to left. This idea need not be confined to memorial church windows; but might with a more secular design be a great beauty in the hall of a country house or in some public building. This dial, to be seen in perfection, should be set in a wall facing due south, though other positions will answer.

Sun-dial in Grounds of Packer Collegiate Institute, Brooklyn, New York.

A curious and interesting memorial may be seen erected on a buttress of Ascot Church, to two members of the Guinness family. It consists of a sun-dial and a lamp, and is intended to indicate the three kinds of light, namely: first, the natural light on the dial; second, the artificial light of the lamp; third, the supernatural light indicated in the inscription, which is —

THOU ART MY LAMP, O LORD.
THE LORD SHALL LIGHTEN MY DARKNESS.

The dial bears the motto: —

I ALSO AM UNDER AUTHORITY.

It shows the frequent irony of good intent that a dial with the same motto and an inscription from Shakespeare, which was erected in Pelton, Somerset, to the memory of a brave soldier, should be torn down and lost, and even the name of the hero forgotten.

A sun-dial has ever been an appropriate memorial gift to a college as well as a useful fitting for a college wall. Many remember a sun-dial on one of the buildings at Harvard University; but it has vanished under "improving" hands. And there was a splendid sun-dial of unusual size and beauty at Williams. Hawthorne writes of the interest he had in it when he saw it at the stone cutter's — its great size and fine finish. That, too, has disappeared within the memory of man. A fine modern dial is upon the Library Tower at Princeton University. Tower and dial are shown facing page 290. Another dial upon a dormitory at Princeton affords a curious mingling of antiquity and modernity; half-timbered walls and a sun-dial are beside electric poles and wires and an American clock. A modern dial on the campus of Yale University is given on page 292. It is of excellent design. Another, on the grounds of

Sun-dial in Grove Street Burying-ground, New Haven, Connecticut.

the Packer Collegiate Institute, Brooklyn, is pictured on page 293. It is a fit and beautiful class memorial. Sun-dials are appropriate teachers in a cemetery. One in a New Haven cemetery is on page 294.

Over the entrance to Mount Auburn Cemetery is a vertical sun-dial, with appropriate and beautiful lines written by the poet John G. Whittier for his friend Dr. Henry J. Bowditch, and inscribed first on a silver dial. Their beauty has made them popular, and they have been frequently copied. They may well close this chapter on the sun-dial as a memorial: —

> WITH WARNING HAND I MARK TIME'S RAPID FLIGHT,
> FROM LIFE'S GLAD MORNING TO ITS SOLEMN NIGHT.
> YET, THROUGH THE DEAR GOD'S LOVE, I ALSO SHOW
> THERE'S LIGHT ABOVE ME BY THE SHADE BELOW.

CHAPTER XIII

PLINY SAIETH: CONCERNING ROSES AND GARLANDS

"Farewell! dear flowers, sweetly your time ye spent,
Fit while ye lived for smell or ornament,
And after death, for cures."

— *Religious Poems.* GEORGE HERBERT.

HOW frequently in reading the pages of Parkinson and Gerarde do we feel a comic resignation at the ever recurrent words, "Pliny saieth," or "Let Pliny tell," "Pliny teacheth," "Pliny reasoneth." We weary of old Pliny's name; but I must add, in truth, we never weary of old Pliny's words. Let us read those words concerning Roses, since they are twenty centuries old, and find what Roses were known and loved in those days of which we are ever eager to learn all details: in those most wonderful and sacred days of the New Testament. Let us turn from the very respectable and easily handled edition of Pliny in the Bohn Library, and take down this vast folio, — companion in size and type and paper and binding of the great *Herballs* of Parkinson and Gerarde and Cole, — cumbersome and heavy to handle, it is true; shedding, too, a leathery brown powder from its ancient calfskin jerkin, yet somehow, a book beloved. Here is the fine ornate title-page, with its beautiful Printer's Mark, a delight to the eye, and the title, *The Historie of the World*, imprinted at London in 1634. It is translated

Roses at Mount Vernon, the Home of George Washington.

by Dr. Philemon Holland, a worthy after Fuller's own heart, and a fit companion for the three Herb-Johns, who were his contemporaries, John Parkinson, John Gerarde, and John Evelyn. Fuller says, "Our Holland had the true knack of translating." The book is too big to hold, but we can prop it up on a reading desk and open at the twenty-first book, which relates to the "Nature of Flowers, and namely those of Chaplets and Guirlands." I will not change a letter of the old spelling, nor a word of the quaint phrasing, since I chose this translation of Holland's chiefly to have the wording of the English of Gerarde's day:—

"Cato in his *Treatise of Gardens* ordained as a necessary point, That they should be planted and inriched with such herbs as might bring forth floures for Coronets and Garlands. And, in very truth, their diversity is such, that impossible it is to decipher and expresse them accordingly. Whereby wee may see, that more easie it was for dame Nature to depaint and adorn the earth with sundrie pictures to beautifie the fields (I say) with all many of colours, by her handyworke, (especially where she hath met with a grounde to her minde, and when she is in a merrie humour and disposed to play and disport herself) than for any man in the world to utter the same by word of mouth.

"To come again to the varieties of flowers; verily there is no painter with all his skill, able sufficiently with his pencil to represent one lively garland of flowers; whether they be plaited and intermedled in maner of nosegaies one with another; or set in ranks and rowes one by another; whether they be knit and twisted cord-wise and in chain-work of one sort of flowers, either to wind and wreath about a chaplet, bias, or in fashion of a circle, or whether they be sorted round into a globe or ball, running one through another, to exhibit goodly sight and entire uniformity of a crosse garland."

By which words it may be plainly seen that though there was great variety and much quaintness of arrangement, yet there was also distinct formality; that set forms were always made, and that each had a signification. Pliny gives an entire chapter to garlands, coronets, chaplets, and nosegays; he tells of their shapes, and why they were called corollæ. He tells that the early Greeks crowned only with leaves and branches of trees, taking no pleasure in "plaiting and broiding of herbes"; they enriched, however, their triumphal crowns with flowers, chiefly

Arch with the Memorial Rose, Twin Oaks, near Washington, D.C.

Roses; and at last Pausias, the cunning painter, and Glycera, the chaplet-maker, started some new modes for Greek dames of distinction and fashion, through as pretty and vivacious a courtship as ancient history can show.

"This Painter was wonderfully enamoured of said Glycera and courted her by all means he could devise. He would seem to counterfeit and represent lively with his pencil in colours what floures whatsoever she wrought and set with her fingers into garlands; and she again strived avie to change and alter her handiwork every day for to drive him to a non-plus at the length, or at leastwise to put him to his shifts; insomuch, as it was a very pleasant and worthie sight to behold of one side the works of Nature in the woman's hand; and on the other side the artificiall cunning of the painter. And verily there are at this day to be seen divers painted tables of his workmanship and namely one picture above the rest entitled Stephanopolocos wherein he painted his sweetheart twisting and braiding coronets and chaplets as her manner was."

After an account of the fashioning of these flower-garlands, Winter Coronets are described gravely: these were made of horn shavings dyed in different colors, "pretty and small," a grievous anti-climax; then came coronets of leaves of latten (a sort of brass) and chaplets of gold and silver spangles; then ribbons followed. All these garlands were hedged about with much formality; they could not be commonly worn. One man carelessly thrust his head out of a gallery window when he happened to be wearing a garland of Roses, and he was promptly carried off to prison.

The grouping of various flowers in a garland had much signification. Beaumont and Fletcher wrote in a charmingly rural measure in *Philaster* of the shepherd, "the trustiest, lovingest, gentlest boy":—

> "A garland lay by him, made by himself
> Of many several flowers bred in the bay;

Stuck in that mystic order, that the rareness
    Delighted me.
Then he took up his garland, and did show
What every flower, as country people hold
Did signify ; and how all, ordered thus,
Express'd his grief : And, to my thoughts, did read
The prettiest lecture of his country art
    That could be wished."

During Rome's magnificence garlands grew in cost and elegance; choicely aromatic leaves were brought from other lands at great expense; chaplets were wrought with needlework and made of silk. There was a formal chaplet of grass, a decoration of honor given by consent of the whole people to some hero; it was sometimes placed on a private soldier with the acclaim of the entire army over his deeds of bravery.

Some years, Thoreau thought, are more directed to the extended observation of nature than are others; at times nature-love seemed to him epidemic, as if all were conscious of the fulness and beauty of life. Certainly this past year is what he called a "year of observation of gardens." We cannot tell how far may extend this lively interest in gardens, and, in sequence, in flowers. We may revive the ceremonial use of flowers which Pliny recorded. The Floralia and Fortinalia of those days had triumphal processions and floral decorations far beyond anything seen by our eyes. We may take lessons and learn to twine garlands; bolder still, we may learn to wear them.

After Pliny has declared that in his day Roses and Violets were the only garland flowers known, he describes the Rose of his time, its manner of growth, and its uses:—

"Roses enter into the composition of sweet ointments and perfumes. Over and besides, the Rose of itself alone as it is hath medicinal virtues, and serveth to many purposes in physick. It goeth into emplastres and collyries or eye-salves, by reason of a certain subtle mordacitie and penetrative quality it hath. Furthermore, many delicate and dainty dishes are served up at table, either covered and bestrewed with Rose leaves, or bedewed and smeared all over with their juice which gives no harm to those viands, but give a commendable taste thereto."

Pliny tells of few Roses, but twelve in all; and gives sadly scant information of each. Johnson, the editor of *Gerarde*, said of Pliny, his book, "Sometimes he is pretty large, otherwhiles so breefe that scarce anything can there be gathered." Brief he is of Roses. He tells of the Rose of Præneste, "the latest Rose," believed to be our Provence Rose; one of Miletus, of a deep and lively red; another late Rose, with never more than twelve petals, thought to be the *Rosa Gallica* of Linnæus; one of Alabanda, "of a baser reckoning, with a weak color turning to white," this perhaps an Eglantine, since it is the "Rose growing in a bramble." Another was the *Rosa Centifolia*, upon which lovely Rose he was surprisingly severe, saying it graced not garlands, save for the extreme ends. Another, the *Græcula*, a large-petalled white Rose, which never opened save when pressed open, has been held to be the *Rosa silvestris*. He speaks of the Damask

Rose Walk at Van Cortlandt Manor, Croton-on-Hudson, New York.

Rose. He gives this curious statement, that sweet-scented Roses ever have the "cup or knob under the floure" (the calyx), "rough and prickly." He fancied, too, that Roses were changed by soil and climate, having more profuse scent in dry soils. This is said to be corroborated by modern observation; but I have found all Eglantines stronger scented in moist soils and places; perhaps the scent of the blossom itself may not be so, but that of the leaf certainly is. He advises cutting, pruning, and *burning* Roses; and he gives a rule to those who "desire Roses to blow early," to dig a trench around the bush and pour in hot water "when the bud beginneth to be knotted."

One point in the culture of Roses, which was insisted upon by Pliny, is just as important to-day; namely, to dig deep in their cultivation, to move the soil at least to the depth of two feet; and Horace speaks of their growing in beds by themselves — a point also clung to by modern Rose-growers, though not by those who love the whole garden more than any single flower.

Pliny asserted that he had scant opinion of trying to concoct certain dishes because they had an influence on the health: he cared not "to mingle Agriculture, Cookery, and Physicke, and thus make a mish-mash and confusion of all things." But he gave thirty-two "searching" remedies to be made from Roses. The use of Roses in medicine is decidedly unromantic and disillusioning. Ashes of Roses "serve to trim the haires of the eyebrows."

"Drie Rose Leaves are of good use in physick; yea the drie Rose cake after the juice and moisture is pressed out of the leaves. Of them be made bags and quilts, yea and drie powders. Wild Rose leaves reduced into a liniment with bear's grease doth wonderfully make hair grow again."

Wine Rosat was thus made, "so saieth Pliny": —

"A weight of 40 denirs (five ounces) of Rose leaves well stamped, put them into a linen cloth together with a little weight that they may settle downwards and not float about. Let them hang thus in 20 sextars (three gallons) and 2 Wine Quarts of Must. Keep the vessel close stopted for 3 Months, then open it and strain the said floures into the Liquor."

It was a belief of Pliny's era, and indeed until the perfected evolution of the botanical system under Linnæus, that a plant with no medical virtue was scarcely worth growing. Botany was for a time forced wholly into the service of medicine. So what Pliny saieth of Roses was simply echoed by Parson Herbert centuries later: —

"A Rose besides its beauty is a cure."

"What is fairer than a Rose,
What is sweeter, but it purgeth."

To both Pliny and Herbert's prosaic and utilitarian notions let me reply in the latter's own words: —

"But I Health, not Physic, choose
Only though I you oppose.
Say that fairly I refuse,
For my Answer — is a Rose."

CHAPTER XIV

ROSA SOLIS, ROSE PLATE, AND ROSEE

"To entreate of them all exactly I doe not entend, for soe a pretty volume of itselfe might be composed. I will therefore give you onely a hint of every one of them and referre the more ample declaration of them to those that would entreate onely of them."

— *Theatricum Botanicum.* JOHN PARKINSON.

"It is not my purpose to make my book a Confectionarie, a Sugar Baker's Furnace, a Gentlewoman's Preserving Pan, nor yet an Apothecaries Shop or Dispensatorie."

— *Great Herball.* JOHN GERARDE, 1556.

QUAINT old Rose recipes for conserves and preserves and their cousinry may be given in this book of Roses without making it a "confectionarie," for they are found in such numbers in so many books of the sixteenth and seventeenth centuries — books that are far from being cook-books. One of my ancient volumes was published in the year 1653, by a noble knight, Sir Hugh Plat, and in the figurative language of the day it is entitled, *A Jewell House of Art and Nature.* It was a book of a class very popular in England at that day and for a century to follow.

In them each author "faithfully set down according to his own experience" all kinds of rare and profitable experiments and inventions in domestic and public life and "Chimical conclusions, rare practices, choice secrets made known,"—in short, as said Sir Hugh, he "anatomized Art and Nature." In all of these books are found ever many recipes for distillations, compoundings, preservings, etc., of fruits and flowers; for such doings were deemed evidences of culture and refinement. Great men and great women aided in compilation. This *Jewell House* abounds in such gastronomical rules; the author waxed somewhat jocose over his turning kitchen-colonel; he writes thus anent the preservation in fresh trim of flowers and fruits:—

"Now methinks I see a whole troop of gallant dames attending with listening ears or longing to learn some new-found skill, as how they may play at Chop-Cherry when cherry-time is past. Wel, to give these Ladies some content I wil unfould a scroul which I had long since as carefully wrapped up as ever any of the Sybels did their fatal phrophiesies, wherein I will make them as cunning as myselfe (having onely that I will reserve one strange venue to foil a scholler withal if need be). The secret is short. Let one element be included in another so the one may have no access nor participate with the other. But this peradventure is too Phylosophical for Women. Then receive it Ladies with plain terms into your open Laps.

"For want of glasses with broad skirts (whereof notwithstanding I do think there are enough to be had if you can be gracious with the Master of the Glass-house) cause new Pewter vessels of large reception to be made of the fashion of bell Saltcellars with divers eyes or hooks hanging in the

Hybrid Sweet Brier Rose.

midst at the which you must fasten the Cherries (or Roses) by the stalks and hang them in it so that one may not touch the other."

Let me conclude his rules. This glass thus hung with Roses was to be set mouth downward into another vessel of water, thus "one element was included in another"; and all this trouble was simply to preserve flowers in freshness of hue, as a rare conceit for the winter season. The knight gave another recipe to preserve dried Roses. I give it at length to show the minuteness of description, the painful ways of our ancestors, as well as their love of Roses: —

"If you would perform the same in Rose-leaves you must, in Rose-time, make choice of such Roses as are neither in bud nor full-blown (for these have the smoothest leaves of all other) which you must especially cull and chuse from the rest. Then take of right Callis sand, and wash the same in some change of waters, and dry it thoroughly well either in an oven or in the sun, and having shallow square or long boxes of five or six inches deep make first an even lay of sand in the bottome, upon the which lay your Rose-leaves, one by one, (so as no one of them touch another) till you have covered all the sand: then with a spoon or with your hand, strew sand upon these leaves until you have thinly covered them all, and then make another lay of Rose-leaves upon the sand, and so make *stratum super stratum* (or four or five lays) one upon another. Set this box abroad in some warm place in a hot sunny day, and commonly in two hot days they will be thoroughly dry: then with your hand, or a spoon you must strive gently to get underneath them and so to lift them up without breaking. Keep these leaves in jar glasses bound about with paper or parchment, in some cupboard that is near a chimney, or stove, least otherwise by the damp of the air they relent again, and so you lose your labour. I find the red Rose-leaf best for this purpose, by reason of his deep colour.

"And so you may have Rose-leaves and other flowers to lay about your basins, windows and court-cupboards all the winter long. Also this skill is very requisite for a good simplifier, because he may dry the leaf of any herb in this manner and lay it, being dry, in his herball, with the simple which it representeth, whereby he may easily learn to know the names of all simples he desireth. The ordinary drying of Rose-leaves is to lay them upon hot leads on a hot sunny day and the sooner you dispatch, the better they will keep their Colour and scent. And when you have dried them thoroughly you may fill a Rose-water glasse therewith, stopping it close and so they will last good a long time."

I love to read these old Rose rules, to note their curious words and details, — that Rose vinegar can be made a perfect ruby color by "making choice of the crimson velvet coloured leaves, clipping away the whites with a pair of sheers" — these were of the Velvet Rose; that "in the pulling of your Roses divide all the blasted leaves and take the other fresh leaves and lay abroad in your windows with clean linen under them." Even the humblest advice comes not amiss, such as "to distill Rose water good-cheap you must buy store of Roses when you find a glut of them in the market, whereby they are sold for sevenpence or eightpence a bushel." Good-cheap indeed was such delectable store at such a price. It assents well with my spirit to read of "kindly ways" of treating the Rose-leaves; while another curious term is a "conceited" mode of preparing Rose-leaves, wherein the word *conceit* is employed in its old Shakespearian meaning of a dainty and fanciful device, especially in cooking. It gives

Roses in California.

me a good notion of the housewife's ample stores to read in these rules of one recipe for "40 bushels of damask Roses," as well to count over the many glass, earthen, and pewter vessels named to make and to contain these Rose treasures. Rules, too, there are for wonderful waxflowers made by coating the real Roses with a preparation of melted wax,—

an almost endless toil; also for casting flowers in glue; another still with "potter's earth,"—a step toward porcelain. More curious still were the flowers moulded in manchet or bread crumbs and fixed with some preparation. A certain noble lord of our own day has been wont to thus divert a dinner party, and cleverly, too. Hugh Plat does not refer to potpourri or any such preparation of Rose-leaves.

"Conserve of red Roses the Italian manner" was defined as a rich paste made by putting the Roses and sugar in a "close stopped glass." Sugar of Roses was made of the deepest red Roses "with the buttons cut off," dried in an oven and "searsed" (which were sifted); these were then mixed with sugar and rolled into sticks. There was a pleasant *Marise*, of Rose-leaves, and pomanders made of Rose-leaves and "Maste of a sweet-apple-tree." These pomanders were carried in the hand or worn in a pierced ball of silver or gold.

> "The bob of gold
> Which a pomander ball doth hold
> This to her side she doth attach
> With gold Crochet or French Pennache."

King Edward gave a recipe for a very good perfume to burn on coals to make the house "as though full of Roses." All these prove why Rose-leaves were purchased in families by the bushel. They were also strewn on the floors of rooms. A distinct item in many ancient domestic accounts was for "strewing Roses and strewing Herbs," and the floors needed them in the abodes of our none-too-tidy forebears.

Rose-leaves were used, too, in large quantities, in manufacturing domestic remedies and medicines. There were Tincture of Roses, Syrup of Roses, Electuary of Roses, Honey of Roses, Rose-troches, and Rose-lohochs (which was a sort of Rose paste to be licked up by the tongue). There was a special cordial made of new Sweetbrier Roses. And Dogroses were made into a pulp and then into a sweet conserve which was called by the ponderous name, "Cynorrhodon."

Every household of any dignity had a Rose-still, and made Rose-water in what would seem to us lavish abundance, and almost extravagance. This Rose-water entered into scores of medical recipes, and was used to flavor all kinds of food, even meats and fish. A Rose-still was a costly cooking-utensil, but one would last for generations.

I have another interesting old book, *The Queen's Closet Opened*, published in 1656, a book of recipes compiled for Queen Henrietta Maria. I purchased it at a book-stall in New York for the sum of ten cents, and I have had amusement from it to at least that amount, I think. Among the prescriptions in the book are some by Sir Walter Raleigh, Sir Kenelm Digby (the viper-husband), the king, the queen, and many physicians and chirurgeons. The part entitled *A Queen's Delight* has many "secrets" for preserving, conserving, and candying; among the materials in high favor were Rose-leaves. All the distillations were done in a Rose-still. I will not copy the recipes for potpourri since they are so well known, but here are some recipes which are not so familiar to us: —

"To make Conserve of Roses boyld. Take a quart of Red Rose Water, a quart of fair water, boyl in the water a pound of red Rose leaves, the whites cut off. The leaves must be boyld very tender, then take three pound of Sugar, and put to it, a pound at a time, and let it boyl a little between every pound, and so put it up in your pots."

"To make Lozenges of Red Roses. Boyl your Sugar to sugar again, then put in your red Roses, being finely beaten and mayd moist with the juice of a Lemon. Let it not boyl after the Roses are in, but pour it upon a Pye plate and cut it into what form you please."

The *Form of Cury* was a roll of English cookery compiled about the year 1390 by the master-cooks of King Richard II. He had in all two thousand cooks. It was a vellum roll containing one hundred and ninety-six recipes, and was presented to Queen Elizabeth as a great curiosity. A few other contemporary collections of cooking recipes exist, and these were all privately printed in 1791 in a large book which is now very rare.

A vast discretion had to be employed by one who followed these recipes. The amounts were seldom given, even of powerful flavorings and ingredients; "according to taste" was the universal rule.

The extreme of vagueness of time-durations in cooking from those old recipes was reached in one given by one ancient Henslow, "Let it seethe the spaces of a mile or more," that is, while you could walk a mile, which meant about twenty minutes. "Seethe it a walm or two" was also far from definite.

Roses, and especially Rose-hips, were of much value to those two thousand royal cooks. Here are two of the Rose recipes from the *Form of Cury*.

"*Rosee* (*from white Roses*).

"Take thyk Rose mylk as to fore welled [before willed]. Cast thereto sugar, a goode porcion pynes [mulberries]. Dates, ymynced canell [cinnamon] and powder gynger, and seethe [boil] it and messe it forth. If thou wilt, in stede of almand mylke take swete cremes of kyne."

"*Sawce Sarzyne* (*Saracen sauce*).

"Take heppes [hips] and make hem clene. Take almands blanched. Frye hem in oil and bray hem in a mortar, with heppes. Draw it up with red wyne, and do therein sugar ynowhg [enough] with powdor-fort [powder of hot spices as pepper, ginger, etc.]. Let it be stondyng [stiff] and alay [mix] it with floer of rys, and color it with alkanet, and messe it forth; and flourish with pomegarnet [pomegranate]. If thou wilt in flesch day seeth capons, and take the brawn, and tese him smal, and do thereto, and make a lico [liquor] of this broth."

We learn many things besides recipes from these old books, among them most ingenious modes of misspelling. But I really believe that *ynowhg* for *enough* is the veriest height of cacography.

When I was a child we always nibbled the hips of Wild Roses and of Eglantine, but had a firm notion that other Rose-hips were poisonous. We called the young shoots of the Wild Rose "Briar-

Rose Arches at Twin Oaks, near Washington, D.C.

candy." I find they were really candied by housewives two hundred years ago.

Rose-hips have been entirely neglected for many

Rose Arches, Twin Oaks, near Washington, D.C.

years as a product for conserves, sweetmeats, etc. I know but one person who gathers them for that purpose, and that is the Irish wife of a German farmer in Old Narragansett. She cooks both Rose-hips and Mushrooms, to the distinctly expressed scorn of many of her American neighbors. From her husband she learned to make the German compote called *Hagenmark*. This seems to preserve the very being of the Wild Rose in its lovely glowing color. When sold in great pails in chill November in the German market-places, the vivid red tempers the frosty air.

Siebold says that the *Rosa rugosa* has been cultivated in China for over a thousand years, and that the ladies of the Chinese court have ever made a delightful potpourri by mixing hips and leaves with musk and camphor.

Here is a recipe for a tart, from a seventeenth century cook-book, entitled *The Accomplisht Cook.*

"*To make a Tart of Hips.*

"Take Hips, cut them and take out the seeds very clean, then wash them, season with sugar, cinnamon and ginger. Close the tart, bake it, ice it, sprinkle sugar, and serve it in."

Pastes of various flowers were made boiled down with Rose-water. Jemelloes were made of sugar, caraways, and Rose-water; these were excellent for "banqueting." "Sugar plate" was similar, save that "gum-dragon" was added. Rose plate was nearly the same thing. Muskechives or Kissing Comfits were made of sugar, "gum-dragon, musk, civet, orris powder and Rose-water," and were "cut into lozenges with your iging-iron." Macaroons were of almond flour flavored with Rose-water. Italian chips were made of flowers of various colors, chiefly Roses. Gingerbread was flavored with Rose-water and gilded, and was deemed a great elegance.

A curious sort of potted Roses was made by the cook of the king of Sicily, and is thus described: —

"This is what I call Potted Roses, and it is thus prepared: I first pound some of the most fragrant Roses in a mortar; then I take the brains of birds and pigs well boiled, and stripped of every particle of meat. I then add the yolks of some eggs, some oil, a little cordial, some pepper and some wine: after having beaten and mixed it well together I throw it in a new pot and place it over a slow but steady fire."

The chronicler adds that when the pot was uncovered the most delicious fragrance issued forth, overcoming the guests with delight.

We have seen that Wine Rosat was known in ancient Rome, and there was a smooth and oily but potent drink of Elizabethan days known by the pretty name of Rosa Solis, strong with *aqua vitæ* and pungent with Orange flower water and cinnamon extract: "old Red Rose water" in plentiful quantity gave it its name.

"We abandon all ale,
And beer that is stale
Rosa solis and damnable hum."

It was beloved of roysterers, scourers, and Mohocks, such rakehelly fellows as Captain Ferrers,

Queen of the Prairie Roses.

who, crazed with this liquor, leaped dare-devil from a high second-story balcony, "the desperatest frolic I did ever see," wrote Pepys. I should have loved to see a braggart gallant of Elizabeth's day, swaggering in bombasted breeches, great green shoe-roses, gauze sash and shoulder knot, and hat with pearl band and feathers, — such a fearful guy as is our poor noble Sir Walter Raleigh in his portrait, — bawling in through a red-lattice for a black-jack, and drinking off in a single vast quaff a draught of damnable hum or "right rosa solis, as ever washed molligrubs out of a moody brain," which was its special function — the cinnamon being deemed in great degree chasing off the megrims.

CHAPTER XV

THE EMBLEM OF THE ROSE IN ENGLISH HISTORY

"Round every flower there gleams a glory,
Bequeathed by antique song or story;
To each old legends give a name
And its peculiar charm proclaim."

"The Rose doth deserve the chiefest and most principall place among all floures whatsoever; being not only esteemed for his beautie, vertues, and his fragrant smell, but also because it is the honour and ornament of our English Sceptre."

— *Greate Herball, 1560.* JOHN GERARDE.

IN that English classic, *Alice in Wonderland*, Alice and her friends, the Mouse, the White Rabbit, the Dodo, and the Lory, all fall into the water. They emerge with difficulty, grievously bedraggled, and stand sadly desponding as to the means of drying themselves. The Mouse bethinks himself briskly: "I'll soon make you dry enough; I'll read you a page of English history; 'tis the driest thing that I know," and he proceeds:—

"The English, who wanted leaders, had been much accustomed to usurpation and conquest. Edwin and Morca, Earls of Mercia and Northumbria, declared for him; and even Stigand found it advisable —"

This regard of English history is common to most of us, but there is a long page of English

Gold of Ophir Roses.

history which is neither dry nor tedious, but is glowing with the richest color of romance; it might be summarized as the Rose in English History. I can but refer to it in a brief regard of the Rose as an Emblem. The "painfull" searcher after facts in regal heraldry finds an indelible record of the Rose in the many badges and devices, especially those of the royal family, which bear a figure of a Rose.

A badge or cognizance was a figure chosen by its owner either as openly significant of some exploit of honor or note of some member of his family; or as alluding to his name or estate or office or calling. A badge was not bound by heralds' rules though it often became an heraldic bearing; it might have been given as a token of the favor of a leader or sovereign, and it was worn openly as a token of allegiance. Each nobleman's badge was prominent in his dress, on sleeve or breast; and it glittered on his standards, his warlike trappings.

Though both were emblems, the device differed from a badge in several ways, the most important being that the badge was an open declaration of the personality of the wearer and its chief object was to accomplish publicity; while the device or imprese had an inner, often a hidden meaning, and was sometimes assumed for the purpose of ingenious mystification. It should not be "so obscure as to require a sphinx to interpret it," Sir William Drummond said, "but should be somewhat retired from the capacity of the vulgar." It must have two parts, a picture and a motto, — a painted, carved, or embroidered metaphor, and also a motto, preferably in a foreign tongue; the "body" and the "spirit" the Italians said. The Italian term *impresa* or *imprese* was used as frequently as the word *device*. We have a lingering bequest of the old mediæval device in our modern book-plate, which should still ever have the "body" and the "spirit." Devices were far more popular on the Continent than in England, and those "curious" English folk who travelled or lived on the Continent — as, for instance, Mary Queen of Scots — were most learned in devices.

Roses at Gravetye Manor, Sussex, England; Home of William Robinson, Author of the *English Flower Garden.*

A few family badges still linger in England, the Pelham buckle being one; and in the history of the English throne we have the badges of the Sun of York, the Broom of the Plantagenets, and the Roses of York and Lancaster; the history of English royal badges would be the feudal history of England.

King Edward I, irreverently called "Longshanks," was the first English sovereign who assumed "A Rose, or stalked proper," as his badge, a golden

Rose in natural form. Many and important were the events of his reign which ended in 1307. It is therefore six centuries since there was a Royal Rose of England.

I shall not enter at any detail into that prolonged story of battle and extinction of the House of Plantagenet, known as The War of the Roses. The first appearance in historic tradition of "the fatal colors of our striving houses"—those of York and of Lancaster—was about 1450. In the Temple Gardens, Somerset and the Earl of Warwick plucked, the former a red Rose, and the earl a white Rose, and called upon every man present to declare himself as to his cause and house. Shakespeare gives a spirited version of the scene in Act II, Part I, of *King Henry VI*:—

"RICHARD PLANTAGENET:

Let him that is a true born gentleman
And stands upon the honour of his birth,
If he suppose that I have pleaded truth,
From off this brier pluck a white Rose with me.

SOMERSET:

Let him that is no coward and no flatterer,
But dare maintain the party of the truth,
Pluck a red Rose from off this thorn with me.

WARWICK:

I love no colours: and without all colour
Of base insinuating flattery
I pluck this white Rose with Plantagenet.

SUFFOLK:

I pluck this red Rose with young Somerset."

Then came, as Shakespeare wrote:—

> "The brawl to-day
> Grown to this faction in the Temple Garden
> Shall send, between the red Rose and the white,
> A thousand souls to death and deadly night."

By some the red Rose is assigned originally to Eleanor of Provence, the queen of Henry III. The tomb of her second son, Edward, Lord of Lancaster, was covered with red Roses. Edward's son was the first Duke of Lancaster, and had on his seal a branch of Roses, and bequeathed to St. Paul's Cathedral his bed, "powdered with roses," which must have been fine indeed — but I wonder where and to what use it was put in the cathedral.

Rose *en Soleil;* emblem of Edward IV of England.

Edward IV placed the white Rose *en soleil* to commemorate his victory at the battle of Mortimer's Cross, where the sun appeared to him "like three suns and suddenly joyned altogether into one," — a singular meteorological phenomenon which is believed to have actually occurred at that time and place.

Another Elizabethan dramatist, Drayton, wrote in his play, *The Miseries of Queen Margarite*, of this strange example of unusual physical forces:—

> "Three suns were seen that instant to appear
> Which soon againe shut up themselves in one.
> So that thereby encouraging his men
> Once more he sets the white Rose up again."

The Rose *en soleil* appears on the Irish groats of this King Edward. An ancient initial with the design of the Rose *en soleil* opens the chapter entitled The Rosicrucians. Edward's favorite badge was a gold collar of suns and Roses with the white Lion of March hanging from it, which must have been a seemly and a pleasing decoration. Edward IV was often called "the Rose of Rouen," he having been born in that town in 1441–42. He presented himself in London when nineteen years old and claimed the English crown. Perhaps the fact that he was "the beautifullest prince of his time" helped his welcome. Agnes Strickland gives one of his coronation songs, which begins thus: —

> "Now is the Rose of Rouen grown to great honour;
> Therefore sing wee everyone y-blessed be that flower.
> I warn ye everyone that ye shall onderstand
> There sprung a Rose in Rouen that opened in England.
> Had not the Rose of Rouen been, all England had been dour;
> Y-blessed be the time God ever spread that flower."

In Edward's reign was produced a beautiful new coin called the *rose-noble.* He issued this coin in honor of a famous victory by sea, thus referred to in an old ballad: —

> "But King Edward made a Seige Royall
> And won the Town, and in speciall
> The Sea was kept, and thereof he was Lord.
> Thus made he Nobles coined of Record."

Sun-dial formerly at The Mount, Astoria, New York, now at Bolton Priory, Pelham Manor, New York; Seat of Frederick Allen, Esq.

He was a distinct believer in magic arts, in alchemy in particular.

Either on this coin, or in one of the reign of Edward I, hangs a tale of magic. Both were exquisitely fine and beautiful, and so much gold was used in the whole coinage that the word quickly spread that it had been produced by the aid of magic. Camden says of the earlier coin:—

"Our alchemists doe affirm as an unwritten verity that the gold thereof was made by multiplication or projection alchemicall of Raymond Lully in the Tower of London.

A rose-noble was held to be a sort of amulet; that the possession of one hindered the theft of a purse containing it.

The antiquary Ashmole gives a circumstantial account of the coming of Lully to England with Cremer, Archbishop of Canterbury, and of a contract with King Edward I to supply him with this coin for which the king pledged war against the Turks. The suspicious king "clapt him up" in the Tower, where the coinage was finally carried out. I have told at some length the story of Lully and King Edward I in my chapter on the Rosicrucians.

By the time of Henry IV so strong was the belief in alchemy that laws were enacted limiting its employment. A serious message exists to a well-known coiner and alchemist, John French, in regard to "p'ctising a true and p'f'table conclusion on cunnyge of transmutacyon of metalls to own p'f't and pleasure." The said French was not to be "letted troubled or vexed of his labour for own p'f't."

Four of the rose-nobles of Edward IV weighed one ounce; there were also double rose-nobles and half rose-nobles. Some of the smaller pieces bore the motto, *Rosa sine spina.* The gold coin of Henry VIII which bore the design of the Tudor Rose with the motto of the rose-noble is, I think, the most beautiful of all English coins. King James had several exquisite coins also, with a Thistie on one side and a Rose on the other.

Emblem of Richard III of England.

Richard III used the badge of a Rose *en soleil*, or a Rose with a Sun; also a falcon with a maiden's head here shown. Sometimes this falcon held a Rose. Another favorite cognizance of the king was a boar, called coarsely by the king's haters a hog. Referring to this badge, a piece of doggerel rhyme was written which caused the beheading of its seditious author:—

> "The Ratte, the Catte, and Lovell our Dogge
> Rule all England under the Hogge."

The king clung boldly and persistently to his badge; for his second coronation he ordered and distributed thirteen thousand cognizances of fustian decorated with boars. He wore it till he died, bearing his standard with a boar—died with "both his legs cut him from"; and then, when dead, was carried "like a hog-calf" hanging across his horse.

Then all the boars speedily disappeared; from standard or sign-board they were pulled down all over the kingdom. A few old inns afterward reinstated the "Blue Boar" or the "White Boar."

With the house of Tudor came in the Tudor Rose:—

> "The rose of snow
> Twined with her blushing foe,"

also a "Hawthorne bush fruited and ensigned with the royal crown proper between the letters H. R." This badge is shown on this page, and the device was chosen to commemorate the hiding of Richard's crown in a Hawthorn bush. It was found while he was being carried off "like a hog-calf," and thrust on his successor's head; this on Crown Hill, which bears the name to this day. Truly those "blended Roses were bought dear."

Emblem of Henry VII of England.

The two Roses, the white and the red of the Tudor Rose, were worn in many ways,—sometimes per pale, sometimes quarterly, usually a white Rose charged on a red one; often they were crowned or *en soleil.* On the marriage of Henry VII Fuller called it "that sweet posie wherein white and red Roses were first tied together."

Henry VIII added a cock, the badge of Wales, to the Rose, and had many other badges and devices. His first wife, Katherine of Aragon, had as a badge a pomegranate open, disclosing the Tudor Rose.

King Henry thus kept Christmas in the eighth year of his reign in honor of Katherine: there was set in the hall of his manor at Greenwich a "garden-artificie" called the "Garden of Esperance." Towered at every corner, and railed with rails of gilt, the banks set with flowers of gold and silver with green satin leaves, this garden centred around a pillar of gold set with precious stones. "And at the top of the six-square pillar was an arch embowered around with gold within which stood a bush of roses red and white, all of silk and gold, and a bush of pomegranates of like stuff."

Emblem of Katherine of Aragon, first Queen of Henry VIII of England.

Anne Boleyn had a complicated device, shown on this page, — a combination of a stump of a tree, a silver falcon with a royal crown and sceptre. From the tree stump came a stiff spray of red and white roses. Her motto was, "To me and mine." No pageant was too extravagant, no honor too great, for King Henry VIII to display at her coronation. Katherine's garden-artifice seems but a poor thing in comparison.

Emblem of Anne Boleyn, second Queen of Henry VIII of England.

Descriptions by eye-witnesses tell of a glory of color and music. There was "a costly and marvellously cunning pageant" on the water. Everywhere was seen the queen's device; on foists, which were lightly built ships, were "mounts bearing a white faulcon crowned upon a roote of golde environed with white roses and red, which was the queen's device, about which mount sat virgins singing and playing melodiously." These foists were strung with streamers, flags, and banners, edged with little "lasserrers," or cords, "hanged with innumerable little bells at the end which made a goodly noyse, and was a goodly sight wavering in the wind." The next day the queen saw a similar exhibition:—

Emblem of Jane Seymour, third Queen of Henry VIII of England.

> "A goodly pageant with a tippe and heavenly Rose and under the tippe was a goodly roote of gold, set on a little Mountain environed with Red Roses and White, out of the tippe came down a faulcon all white, and set upon the roote; and incontinently came down an angel with great melodie, and set a close crown of gold on the faulcon's head; and in the same sat St. Ann, with all her issue; and under Mary Cleophe sate her four children, of the which children one made a goodly ovation to the Queen."

Fountains of wine and conduits of sweet water ran everywhere; cupboards of silver, displays of gems;

angels and graces vied in singing praises; hippocras and wafers were significantly partaken. All the clergy glorified her, all the duchesses and countesses bowed before her, and there was not one who knew of that other terrible decrowning so soon to follow.

Poor Anne of Cleves had a fresh device, and the fatuous inscription on her wedding-ring was, "God send me wel to kepe." Katherine Howard and Katherine Parr both had augmentation of arms, and an ugly device of a woman's head crowned, surmounting a segment of a triple Rose, which is here shown. Jane Seymour's motto was, "Bound to obey and serve," and a singular device of a castle, tree, phœnix, and Tudor Rose. It is shown on page 329.

Emblem of Katherine Parr, sixth Queen of Henry VIII of England.

Henry VIII's son Edward did not use the Rose in his cognizance, but Queen Mary had a singular badge, shown on page 331, composed from those of her parents. In it appeared the Tudor Rose, a sheaf of arrows, and a crown. Queen Elizabeth had for her badge a Rose crowned, for England; a Fleur de Lis crowned, for France; a harp ensigned with a crown, for Ireland. Many of the coins of her reign bore the figure of a Rose. "Bright-red Rose without a Thorn" is on some of her coins. I suppose no queen ever lived who more fancied complicated allegories, emblems, symbols, and devices; they were an unconscious revelation, an evidence of the slyness of her nature.

James I had as a motto, "Blessed are the peaceful." With him the Scotch Thistle entered the royal badge ever to remain; and its representation on page 332 proves his badge a very pretty one.

On the coins struck for the coronation of Charles I, 1633, was a great Thistle with the motto: "Here grow our Roses." Queen Anne had on her seal a Rose and Thistle springing from the same stem, and the motto, *Concordes*. The present royal badges were settled in 1801, and that of England is a white Rose within the red Rose. Thus has the Rose felt the very heart-beat of English history.

Emblem of Queen Mary of England.

May we not, then, in the recollection of all this allied Rose history, glow with the pleasure of retrospection allied to present gratification in the sight of a beautiful York and Lancaster Rose? This storied Rose has been pushed aside for many years by the hybrid perpetual Roses; but now that the love for old-fashioned flowers has risen with such force, it is again offered for sale, and promises to have much popularity. It is a brave creature, having a clean-cut, bold striping, and mingling of pure white and bright red. It is sturdy, too, in growth, not a clinging, gentle flower, but one fit to be associated with the history of wars.

Emblem of James I of England.

The Rose we here call York and Lancaster was called *Rose versicolor* by Parkinson in his *Paradius in Sole, Paradisus Terrestris* — A Garden of all Sorts of Pleasant Flowers. Mrs. Ewing founded a Parkinson Society to promote a love of old-fashioned flowers; but Parkinson's book is too rare to be of influence here. When I inquired throughout our public libraries in 1901, not a copy was to be found in America. Fortunately I secured for my daughter's collection of old herbals and flower-books, begun when she was a little girl, a copy of Parkinson's *Paradisus in Sole* in the first edition, and the constant reference to it and Gerarde's *Herball* (the second edition) have been an infinite pleasure to me. I would I could quote Parkinson's words in full upon the York and Lancaster Rose!

CHAPTER XVI

OUR GRANDMOTHERS' ROSES

"To be chronicled, and chronicled, and cut and chronicled, and all-to-be-praised." — *Worthies of England.* THOMAS FULLER.

TWO Rose-books are to me absolute authority for Rose-knowledge of one hundred years ago, ere the Rose-treasures of China were lavished on our gardens. One is Miller's *Gardener's and Botanist's Dictionary*, London, 1807. My four volumes are of great weight, literally, being nearly nineteen inches in height. I think this vast size and weight has kept them from frequent consultation, hence the perfect condition of the lovely binding. It is a splendid work, so temperate and lucid in description; in it forty species of Roses are given with the different varieties of each — about two hundred in all.

The second book is beloved of all Rose-lovers, *Les Roses* par P. J. Redouté. Two of the illustrations are reproduced, but colorless, on pages 353 and 354. At first sight of the exquisite Roses which bloom on these pages they seem to be hand colored by some most skilful, sympathetic, albeit old-fashioned water-colorist. But they are not tinted by hand; they are wonderful examples of what was then a newly discovered process of color-printing. The book was issued in 1824, fast approaching a century ago, and is unexcelled by any Rose-book to-day.

The Empress Josephine had a charming Rose garden at Malmaison; and when the Allies entered Paris in 1815, vigorous orders were given to protect always her garden. I am told there still exists in the English navy a standing order since the time of England's war with France, that all seeds and plants and shrubs bearing her address, when seized on French vessels by the English, should be at once forwarded to her, a true courtesy of war.

Though the Rose is the emblem of England, many of our best Roses bear French names; for no marked progress in Rose cultivation in any country took place till 1815, when Vibert, the first of the great French Rose growers, founded his nursery. It was owing to the patronage of the Empress Josephine that he could accomplish his great results. It is only forty years since English Rose growers raised any important Roses. In imitation of Jacquin, Redouté brought out in 1803 a sumptuous book on this garden at Malmaison, dedicated, of course, to the empress. He also illustrated Rousseau's *Botany* with sixty-five beautiful flower-pictures. They had charming Roses in 1824 — Roses which would not all take prizes at Rose shows to-day, but very cheerful flowers, wholesome, hardy flowers, and some of them very delicate and exquisite flowers.

I have had a great deal of happiness this week in going through the pages of Redouté in company with Miss Jekyll's and Mr. Mawley's *Roses for English Gardens*, which has just come to me. A photograph does not convey a good idea of the personality of a Rose; with the distinguishing color gone, a curious sameness appears. I have seen an experienced Rosarian sit puzzling over a photograph of a Rose-bush, where the leaves and stems were not very plainly shown, and the pictured flowers were small, hesitating over the correct naming of the photographed Rose. And worse still, but amusing, a lover of Roses had last summer about forty-five photographs taken of Roses intending them for lantern-slides to use to illustrate a carefully studied lecture on Garden Roses which he was to deliver before a very intellectual audience. Three or four weeks elapsed ere the pictures came to him; and then, to his chagrin,

June Rose.

he found he could not use half of them; for when there was no tell-tale attribute, such as a neighboring stick, a known grouping, or some unusual distinction of stem or foliage, as has the Burnet-leaved Rose, he could not positively name his Roses.

I wonder if this simple picture of a Rose in bloom (page 335) will convey to any gentle reader who glances at it in the winter months when this book will be seen by that reader's eyes, any of the pleasure of reminiscence of the old-time garden which the photograph brought to me when I first saw it on a bitter day in March. I had not in it the carmine color, but I had the precise shape of a Rose I loved, the first Rose of summer, the June Rose, which might almost be the May Rose, since it has laughed outright with its cheerful glowing blossoms, and become silent and grown all green again before any other June Rose has opened a bud. It blossomed in an interval of time — a week only — when nothing else bloomed, save possibly Money in both Pockets. Thus it had a double welcome, — it was for a time the only Rose, and it was the only flower. This picture does not, perhaps, give to one who had never seen this Rose a notion of the saucer-like expanse of the full-blown petals when in the hot sunshine. Its pure crimson disc is a bit like the *Rosa rugosa*, but a more perfect circle and an infinitely better color. This old favorite still is seen in many front yards in New Hampshire; at one farmhouse it has filled the entire yard. It is called there the Hedgehog Rose, from its sharp thorns. Its fragrance is precisely that of an American Beauty Rose;

and it is a close-grained, compact fragrance,—a fragrance of weight. You can almost imagine a little globule of solid perfume standing on each crimson

Scotch Roses.

saucer ready for you to gather up as you pass, and keep forever as a memorial of spring and sunshine.

There is a May Rose of the botanies, *Rosa majalis*, a native of Sweden and Lapland, and of Yorkshire, said to have "pale red" flowers. It is supposed to be the single state of the *Rosa Cinnamonea*, and from other details of the description might be my

June Rose, if you could possibly conceive any one calling the color a pale red. But there were strange vagaries of color-naming and seeing in the old botanists.

The next Roses in bloom were the Scotch Roses, yellow and white; these seem little cared for to-day. They do not, of course, satisfy any ambitious Rosarian. Their peculiarities of growth are well shown on page 337. Yellow Roses were so few that the yellow Scotch Rose held in favor longer than the white variety. The Austrian Yellow Rose, centuries old, and the Persian Yellow are little seen; the latter is such a vile renegade as to its scent that it should not be called a Rose. It is an unutterable Joan Silverpin, "faire without, foule within," and owes its uprootal from many a garden because it "smells to heaven." The Harrison Rose is by tradition named to commemorate a political event, the election of the first President Harrison.

We never gathered the yellow Roses, for two reason: their frail petals dropped so quickly that they were valueless, and their spiny armament was too defensive for ordinary attack. It seemed to show that they were never intended to be picked. As for the Persian Yellow—there was a third reason already told. Long straggling branches, all closely armed, carried the yellow bloom of the Scotch Roses afar, and gave a rather shapeless appearance to the bush, one that would be abhorred by modern lovers of Rose-pillars and Rose-standards.

Stretching out thus in each branch, with its finely cut, close-set leafage, it certainly had the effect of a close-tied wreath, with yellow blossoms tied in, short-stemmed, to the very end of the drooping, swaying wreath. Country folk called it the Yellow Wreath

Harrison Yellow Rose.

Rose. Whatever the scent of the flower of the Harrison Rose, and all Rose friends differ, even to the extent of questioning whether it has any scent, the leaf certainly has the true sweet-brier fragrance.

Yellow Wreath Rose.

Only a whiff, however, which you cannot perceive at all when you try to; nor could any one else smell it if you told them of it.

Mrs. Martha McCulloch Williams calls the Burnet-leaved Rose the Burr Rose, and its calyx is covered with close green prickles like a burr; she compares it to a "flat dish of crinkly pink crêpe," which is a fine description.

There was a nameless Rose of rampant and cheerful growth, a Rose which was everywhere. Its purply crimson deeply cupped blooms were fine in shape and color, not being magenta-tinted when in half-bloom; but when fully expanded or when picked they turned a dulled tint, looked faded, muddy-colored, and showed white streaks at the base of the petals. You always found this Rose in the poorest homes, usually at the kitchen door, where it fairly kept down the weeds. We called it the Purple Rose; I think it must be the Purple Bouquet of Mrs. Williams, of which she says:—

> "In all the Rose Kingdom there is no other burgess so cheery, so thrifty, withal so happy as the Purple Bouquet. It asks only leave to grow, never a chance; will spread to a green mound or shrink to one starveling stalk, as fate may ordain. Great or small, it has always bloom and to spare for you."

This Purple Bouquet was one of the family *Rosa gallica* or Provins Rose, and I think the *Gallica officinalis;* I hope I shall not seem "to say an undisputed thing in such a solemn way" if I remind my reader that Provins and Provence Rose are very different garden Roses; the Provence is the *Rosa centifolia*, the French call it always *Rose a cent-feuilles.*

I know not the precise name of this Purple Bouquet, for nearly all of this family are purple-flushing. Redouté gives several which might be our Purple Bouquet, and the common Provins Rose, known as the Apothecaries' Rose, is in the illustration precisely like our Rose. One, the type, the nearly single *Rosa pumila* or Austrian Rose, is the *Rosier d'amour* of the French book, though that is not the Rose so known in England. Another was called by the Frenchman the "Cabbage Provins."

One called by Redouté the Purple Velvet Rose, I

recognize as our Black Rose. This was so color-flushing that even the wood had purplish black streaks. This and several of the fine Roses of this family were grown for the garden of Josephine at Malmaison, and the finest propagated by themselves when the garden was comparatively neglected, after her death; they were sturdy, independent things; you couldn't kill them; underground roots carried them far and wide. I have seen a backyard and vegetable garden, the kitchen yard of a deserted farm-house, over one acre in extent, covered with a mass of these Purple Bouquet Roses, which had triumphed over every other growing green thing. And I have seen a village sidewalk bordered for a quarter of a mile with a natural Rose hedge of this variety. I recall that one New England rosarian — I call her that though her whole garden wasn't more than a hundred feet square — always called the purple-pink Rose of her garden the Sultana Rose. This had but a double row of petals, and when grown in the genial sunshine, which seemed ever to shine on this garden of cheerfulness, these petals were of an infinitely rich color. I was delighted to find in Redouté Mrs. Pyncheon's Sultana Rose, under the name *Belle Sultane*.

Few of the other Provins Roses are grown in the gardens of great Rose growers, and these few are not our Purple and Black Roses. Curiously enough, the poorest to my mind of all, the Striped Provins, still are grown in England, among them Redouté's *Gros Provins Panachées* and *Gallica versicolor;* this is called by Redouté the Rosamond.

The *Rosa lucida* is the Rose called in England by the pretty name *Rose d'amour*. This is a true American Rose, a native. I have seen it growing this summer many times, and a very cheerful bush it is, with its shiny, glossy leaves turning a gay light yellow, with some crimson red in autumn, and bearing many odd flattened hips. In gardens its flower is sometimes double and sometimes single, with a few inner half-rays of petals within the perfect row. It is a very different pink from the Provence Rose or Cabbage Rose, — a redder pink, — and has not the beautiful fragrance; in fact, it is one of the few Roses whose scent is distinctly distasteful.

Baltimore Belle.

The only truly American roses seen universally in our gardens are the beautiful climbing Prairie

Roses (*Rosa rubifolia*). These were originally a native of Michigan and other Western states; and the clear pink single variety, known as the Michigan Rose, still is grown, and an arch of it is a perfect thing. About 1836 the Feast brothers, florists in Baltimore, developed this Rose, and gave to us the beautiful Baltimore Belle, the Queen of the Prairie, Anna Maria, and Gem of the Prairies, which last is slightly scented sometimes, — not always, I find.

These all have large rough dark-green leaves of five to seven leaflets. They are the hardiest climbers known, and are far more rapid in growth than the Ayrshire Roses. And they come, too, when many other summer Roses are gone, when the Hundred-leaved Rose and the York and Lancaster are passing. They are not so delicate as the climbing Tea-Roses, but they are more generous; one characteristic is their lavish fulness; they seem fairly crowded with petals. And another is their wholesomeness; they are fresh with the primeval breath of the Michigan forests and fields.

When they could have the Baltimore Belle, I cannot understand why any one planted the Boursault Roses, but they grew in every dooryard, always by the kitchen end of the farm-house. Forty years ago there was scarce a woodshed in New England but was garlanded with the crimson Boursault Rose. It was as widely planted in its day as the Crimson Rambler in our own. I care little for any of the Boursaults, but the crimson variety was certainly cheerful. They were firm climbers and almost thornless. What wonderful things were the new stalks of the Boursault!

Massive, straight, strong, they pushed like Zedekiah. Tremendously big-looking they were in proportion to the blossom, which seemed dull of tint beside the rich dark purples and reds of the young stalks with their curious plumlike bloom.

Anne de Diesbach Rose.

At the base of page 345 is shown the lavish growth of the Anne de Diesbach Rose, one of the "middle-aged" Roses which have been deservedly popular for a score or two of years. This bush is twenty years old, — the hardiest, sturdiest thing, bearing every trying extreme of New England winters, and every trying pest of New England summers, and sending out each year its many hard, round buds (the *knops* of Chaucer), which should be gathered when half open if you wish to secure within doors the rich, luscious pink of the outdoor bloom. Another beautiful "middle-aged" Rose, the Lawson Rose, is wonderfully satisfying. One of the first of its plants was set in the Manning garden, in Salem, where all flowers prosper, and, above all, all Roses. Its pink is not like that of the Anne de Diesbach, but is equally beautiful.

There was a White Rose of the old garden which was to us nameless, a June Rose of fullest bloom, and greatly beloved. We called it simply "The White Rose," never confusing it with the Madame Plantier. It sent out very long, strong shoots, with strong, small thorns and rather bluish green leaves, which were easily ravaged by Rose enemies. There were two or three, or even five, Roses in a group together — not in a thick cluster, like the Seven Sisters — but always with one Rose a bit bigger and finer than others of the group. This flower was very rich and full, perfectly double; the others might show a trace of gold in the heart. Every freshly opened Rose had a faint suggestion of pink; but on the second day were absolutely purest white,

The White Rose.

and when they dropped, almost blue-white. Occasionally a single lovely snow-white Rose would appear in bloom on this bush in the autumn, and was a great prize, to be carefully gathered and set in the slender, wrought-silver vase. It used to be in all gardens, and was honored in all gardens; but I know but one garden in Worcester where there is a large and strong Rose-bush of "The White Rose." It is shown facing page 346. I know not of what Rose Cowper wrote, but his lines might well be written of The White Rose, as it stood against the "darkest gloom" of a tall cedar. He says: —

> "The scentless and the scented Rose; this red
> And of an humbler growth, the other tall
> And throwing up into the darkest gloom
> Of neighbouring cypress or more sable yew
> Her silver globes, light as the foamy surf
> That the wind severs from the broken wave."

The White Rose was, without doubt, the *Rosa alba*, for its leaves had the bluish tint of that race. It is of questioned standing in Rose lists, being, apparently, a cross between the *Canina* and *Gallica*. It had a cousin, with the same glaucous leaves, which leaves had an almost artificial appearance. This cousin was the Blush Rose, or the Maiden's Blush, deemed by many old-time Rose lovers to possess the most delicate and exquisite tint of any Rose. This Rose was easily blighted as to its blossom and its growth, but was valued all the more because of its frailty. It was absent from our garden for years, but we have now a Blush Rose-bush, which we found in an old garden. The story of its finding has been told so perfectly in verses by Edgar Fawcett that I will give them instead of my own prose: —

"I lean across the sagging gate;
In rough neglect the garden lies,
Disfeatured and disconsolate,
Below these halcyon skies.

"O'er pleasant ways once trimly kept
And blossoming fair at either verge,
Weeds in rank opulence have swept
Their green annulling surge.

"But over there, as though in soft
Memory of bloom that no more blows,
A Rose-bush rears one bough aloft
Starred with one stainless Rose.

"Above these weeds, whose ruffian power
So coarsely envies what is fair,
She bends her lightsome dainty flower
With such patrician air

"That while I watch this chaste young Rose
Some pale, scared queen she seems to be,
Across whose palace courtyard flows
The dark mob like a sea."

The Madame Plantier, "the other White Rose," as we called it, has never wavered in popularity nor waned in goodness since the first June when it opened its eyes on American soil. It was warmly welcomed, and has been ever, in turn, a devoted citizen. Other Roses often give to us on occasional years but a handsel of bloom; but the Madame Plantier ever lavishes upon us, whether in formal garden or cottage border, whether the season be cold and backward or dried up with sudden drought, a profusion of snowy blooms, as close-gathered as the stars in the Milky Way.

The Damask Rose has ever been a favorite comparison to indicate the delicate complexion of a fair woman.

A pretty name may be given the lovely Damask Rose. An old Dutch carol runs thus: —

"My master hath a Garden,
Which flowers fair adorn,
And lovely Damask Roses
Are there called Patience."

Patience.

Old Aubrey wrote nearly three centuries ago of that fair creature, Venetia Stanley, wife of Sir Kenelm Digby:—

> "She had a most lovely sweet-turn'd face, delicate darke brown haire. She had a perfectly healthy constitution; strong, good skin; well proportioned; enclining to a Bona Roba. Her face, a short oval; darke-browne eie-browe about wch much sweetness, as alsoe in the openinge of her eie-lidds. The colour of her cheekes was just that of the Damaske Rose, which is neether too hot nor too pale."

This certainly is a beautiful drawing of the wife who died suddenly and was reputed to be poisoned: "wch her husband imputed to her drinking of viper-wine; but spiteful women would say 'twas a viper-husband." And I always recall this term *viper-husband* when I read Digby's pious passages. A handsome portrait of her still shows her damask cheek. A beautiful old Damask Rose-bush over one hundred years old is pictured at the base of page 349.

The darker shade of the Damask Rose was called the Velvet Rose; it is now seldom seen save in oldest gardens, but I shall ever love it for its richness of color of leaf and flower, so rich that color seems fairly to ooze from the entire plant. A single Velvet Rose in a vase within doors is such a point of color that it dims all else in the room. It is the finest red-pink in the whole Rose world,—nay, more, in the whole flower world.

The Velvet Rose is not graceful in bud,—it has then a blunted look; nor is the shape when fully opened the most elegant of Rose forms; but the texture! the color! the scent! to tell which of the three is the finest attribute puzzles the Rose lover, when all are perfect. I have a splendid Velvet Rose given to me by a rare old lady of Maine who calls it the French Rose. Well is it named, since it is the "living emblem and the sign" of a wonderful romance of French history. This Rose tells the story of the "Affair of the Pinks,"—a romance which is crying out to be told in an historical novel by some gifted hand. There was but one perfect bloom this year on the old Rose-bush which was transplanted from the garden of Fountain la Val. Such a tiny, aged, broken thing, its leaves are fairly gray with years and the stem is lumped and seamed and wenned, yet here is this sightly, this luscious bloom. Its first American home was at the La Val garden at Lamoine on the Maine coast, northeast of Mount Desert. Ancient fruit trees and Lombardy poplars mark the site of the French mansion which was the home of Madame la Val, a French aristocrat, a widow of the Revolution, who came with her daughter and thirty citizens to found a refuge for her distressed fellow-countrymen.

The story of this house and the proposed refuge there of Marie Antoinette is part of the "Affair of the Pinks." The picture of the beautiful French queen on the brig *Sally* of Wiscasset, Captain Clough, is to me a curious one. The story has the attendant figures of Mirabeau, Count de Fersen, Talleyrand, Lafayette, in sharp contrast with the Yankee skipper and stately Madame Swan of Dorchester, and fat General Knox. The cargo of rich goods came without the royal owner to Squam Point. I have seen fans, china, silver, which were part of that cargo. In my sister's book, *Furniture of the Olden Time*, is shown a splendid semicircular sideboard, with knife-box, great silver salver, and urn, now owned by Hon. John P. Baxter of Portland, Maine, and known as the "Marie Antoinette Side board." The furniture in that ship's cargo went largely, I believe, to the Swan mansion in Dorchester. The son of General Swan married the daughter of General Knox, and the sideboard was part of her wedding furnishings. From the Knox mansion it came to Mr. Baxter. Truly this French Rose blooms full of wondrous interest. In the writings of Carlyle, of Sewall, of Gouverneur Morris, of Lafayette, of Madame de Staël, of Talleyrand, do we learn of the "Affair of the Pinks." But we would wish that Morris had not been so guarded in his diary notes, nor Lafayette so silent when he visited Madame Swan in 1824.

We had in our garden Bourbon Roses, Bordeaux Roses, Burgundy Roses, and Boursault Roses, — names easily confused in village communities. I found in one town the Boursault Rose called by every one the Bourbon.

The Bourbon Rose was much esteemed, and its beautiful late blooms in the autumn were all the greater delight because these blooms were so few. This was discovered originally on the Isle of Bourbon growing in a hedge of Bengal and Damask Perpetual Roses, and it has characteristics of both, yet is absolutely a distinct variety; some folk called our beautiful great pink Bourbon Rose what I fancied in childhood was spelled Apple-een. I now know it was Appoline, a Rose introduced here in 1848. Our Bordeaux Rose was much like the Burgundy, save that the latter was smaller and dark red, while the Bordeaux was pink. I have found in Redouté a perfect representation of this Bordeaux Rose. Redouté says it differs little in general shape from, and often in rich earth will grow as large as, the Hundred-leaved Rosc.

Bourbon Rose from Redouté.

I have given on page 354 a reproduction of Redouté's drawing of the Burgundy Rose. I cannot give his beautiful coloring, alas! which is peculiarly happy in this case, giving the exact tint of clear redness. We called it the Little Burgundy, and the name always seemed to me so appropriate for a Rose of that color; while another Burgundy of our garden — white with a beautiful pink centre — should have had another name. It did have, in fact; it was the *Pomponia Tudor*, while our Little Burgundy was the *Pomponia Burgundiaca*. The white Burgundy changed color very decidedly in the course of its bloom. All the petals were pink and the edges and outside petals turned white as they opened. Redouté gives the color of the Little Burgundy as a "*rouge pourpre foncé*," which is a very good description. He says it should be cut immediately after flowering; and I recall distinctly that we always cut the withered blooms.

Little Burgundy Rose. Redouté.

The beautiful Bengal or China Roses I never knew in our garden save in one form, the tiny Fairy or Pony Rose. This has ever been to me one of the fully satisfying and more than satisfying things of the old-time garden. Some ever find beauty in vastness; they prefer ample extent and forms in all things; they love large, full Roses. I have ever inclined to love beauty in miniature, finest lace and drawings, bits of carving in coral and ivory, clusters of perfect tiny gems rather than a single large one; so I love small roses. I should love a miniature Rose garden set with Pompon Roses. Even as a child I loved especially this Pompon form,—a Double Buttercup, a Flowering Almond or Double

Cherry, an English Daisy; and in Roses this Fairy Rose, and the Dwarf Burgundy Rose, the White Pet, and the Banksias. I am sensible of some attraction, some drawing, toward these Fairy Roses which I can scarcely explain. They are not appealing through their tiny size nor have they any clinging frailty; they are compact, vigorous, wholesome; they have such a confident and cheerful expression; they thereby assume an independence and dignity which comes to any created thing of any size which plays its small part in life to perfection.

Fairy Rose.

I note with some distress that my favorite writer on the Rose dismisses the Fairy Rose with these curt words, "We do not deem them of value, the Bengals are small enough;" nor does he place the Pompon Roses on his lists. This is not, of course, through a dislike for the double form of flowers, which many feel for certain species. I never like to see a flower of irregular or complex outline doubled: a Pea, a Columbine. But Roses, Pinks, and Peonies have, by doubling soft petal on soft petal, secured clustered beauty instead of outlined beauty; as have all our fruit trees, the Peach, Almond, Cherry, Plum, Crab-apple, whose exquisite doubled flower forms have all the charm of the double Rose. One thing can be said in favor of doubled flowers in general, — namely, their scent is doubled with their petals.

There is one Rose of our grandmother's garden to which is constantly imputed a false old age. Especially in the South, where it grows freely, will great age be assigned to the Seven Sisters Rose. I have often been told that a certain Rose of this variety was planted long before the Revolutionary War; or that it is one hundred and fifty years or even two centuries old. You must never question such statements; it is cruel to the Rose owner, and besides you get yourself disliked — and for an unimportant thing. The Seven Sisters Rose was, in reality, brought to us from China in 1821 — and that is old enough for anything or anybody.

The Seven Sisters Rose is given many names. It is really one of the forms of the Japanese *Rosa multiflora*, and was known as the *Rosa Roxburghein* and finally as the *Rosa Grevillei*. A good description of it is found in Loudon's *Arb. et Fruct. Brit.*, where it is called "*Rosa multiflore Grevillæ* or Seven Sisters Rose." As the *Grevillei* it is offered in our America catalogues, though the Rose pedlers, who have sold it in such numbers, always call it the Seven Sisters.

I find that the Seven Sisters was wonderfully praised when new to English and American folk.

Seven Sisters Rose.

It had a popularity like the Crimson Rambler. There were pink and white and purple-pink varieties; and a story is told of a plant at the Holdsworth Nursery which in 1826 covered one hundred square feet, had over one hundred corymbs of blossoms, and some of the corymbs held fifty buds. Some thought that the variety of colors in the buds — seven in all — gave it its name.

In 1830 Lindley wrote of it: —

"The Chinese call it the Seven Sisters Rose because about seven flowers open at the same time, each varying from the other in tint."

One good quality of this Rose is that the blooms are exceptionally long-lasting, and they are a neatly shaped Rose; white, at first opening, but tingeing with pink after a few days; and they have a delicious fragrance, — all these traits make them charming Roses for vases. It is never recommended by Rosarians, for it is held to be so free a grower and flowerer that it exhausts itself, seldom living but a few years.

A gazel or ghazal is a form of Persian love verse. Each couplet is, in a sense, a perfect poem or thought, but there should be thirteen stanzas in all, and the first and second lines rhyme with the fourth, sixth, eighth, and so on to the end. The ghazal is in sentiment something like the Chinese "stop-short"; the thought of the couplet should be carried on beyond the expressed words. Hafiz, the Persian poet of the fourteenth century, wrote this ghazal upon the Hundred-leaved Rose: —

"Thou fairest Rose of all, ah, say
For whom dost thou thy hundred leaves display?
To what blest mortal wilt thou own
Such charms have sprung for him alone?"

The poems of Hafiz might well be entitled Hundred-leaved Roses, for his pages are filled with crowding Rose-leaves of verse.

The Rose of a Hundred Leaves is thus sung by Moore in *Lalla Rookh*: —

> "The joyous time — when pleasures pour
> Profusely round, and in their shower
> Hearts open like the season's Rose —
> The Floweret of a Hundred Leaves,
> Expanding while the dew-fall flows,
> And every leaf its balm receives."

Although the Hundred-leaved Rose is a flower of aristocratic lineage, it is most countrified in bearing. It has a simple countenance, almost a childish look, and its pink and white is like a country child in good health; I always think of Shakespeare's phrase, "a shining morning face," when I look at it. There is in this Rose much of "the freshness of the early world"; it speaks of the childhood of humanity.

The York and Lancaster Rose bourgeoned in every garden of any grandmother of English lineage. I have told its story at some length in my chapter upon the Rose in English History, and so will only name it here to assure it of its ever welcome presence in the old-fashioned garden.

Moss Roses were known to Gerarde and Parkinson. Ever subject to that most trying of Rose scourges, mildew, they would have been crowded long ago from our gardens had it not been for a certain tender sentiment they awaken in every one. Many fanciful verses have been written by poets of many lands to give a sentimental reason for the mossy greenery of this Rose. The moss is in no sense incongruous, yet it has an element of the unexpected. To many it gives the highest point of

Hundred-leaved Rose.

charm and beauty to the Rose. I can scarcely enter into this feeling; though as a child I know a gift of a Moss Rose was an act of unspeakable sentiment. Infinite in number have been the Moss Roses since Shakespeare's day, but still is the Moss Rose, the common pink Moss Rose, the *Gracilis*, the best of all its kind.

We have seen the few Roses of Pliny's day and even of the times of the War of the Roses; our grandmothers had more to choose from, but after all they had a short Rose-list. Nor had they Rose gardens. Their Roses grew among other plants. Often a single bed would be given to Roses, particularly those which needed special treatment; but it would be a small bed. Our grandmothers' gardens never presented the expanse of tall thickly set sticks and scant low growing foliage too often seen in a Rose garden, and suggesting Hogarth's famous drawing on a ticket for a flower-show of a monkey watering a row of leafless sticks. Why, I have a photograph of George Bancroft's Rose garden which I simply could not persuade myself to display in this book. I looked at it a score of times and then thrust it in a remote writing-desk; I am glad now I did not show to any one that monstrous stretch of well-kept lawn in the foreground with two long beds of unvarying tying sticks and scarce a bit of foliage or bloom, standing well up against the background of tall hedges. For I saw that garden when I was a child. I recall distinctly that I wearied over the long stroll down the path, and the interminable and eager discussion over certain Rose plants which seemed to me scarce more than Hogarth's exhibition sticks, and not worth any glance, much less any animated talk. And I longed to go home, when suddenly we came upon the Tea-Roses — the finest in this country, the first I had ever seen in any number. I have never forgotten them. It seems to me that I had never smelt Tea-Roses until that afternoon, the beauty and scent of those wonderful Tea-Roses is imprinted forever in my brain.

In some vicinities the Tea-Roses acclimated themselves to a wonder. Martha McCulloch Williams wrote to me in a letter of one Tea-Rose — known as the *Triomphe de Luxembourg* — in her early Tennessee home, where the land was not far from virgin soil, and where thrips, red spider, rose-chafer, and lady-bug were entirely unknown: —

"The Luxembourg is the Queen of Tea-Roses to my mind, much the shape of the Catherine Mermet, but richer, and of finer foliage, and colored beyond all other Roses, not even excepting *Gloire de Dijon*. It is hardy in Tennessee; here in New York I dare say it would not live out. Old bushes give the best flowers, and those which opened on the very edge of the forest were always finest of all. The heart was variable but generally creamy pink with golden suffusions. The outer petals run between deep red bronze and copper-yellow. It was, I think, one of the earliest high-colored Teas perfected. It grew rampantly, and had so much red blood that the leaves and flowers at first unfolding were almost as high-colored as the flowers. It is my Rose of Roses.

One great question must be ever in a Rose garden, — whether to plant only Roses, or to cover the ground with some low-blossoming plant, or some greenery? One friend has planted under every Rose-bush blue Pansies. Some were very pale blue, some a cheerful dark blue, but a pure sapphire color prevailed; as the Roses in bloom when I visited

Rose Garden at Mount Vernon-on-the-Potomac; the Home of George Washington.

this garden were nearly all yellow tinted Tea-Roses, this garden seemed to me in perfect taste. My sister has the best "cover-ground" I know in the first year's growth of Adlumia, what we always called Virgin's Bower. It is as graceful as Maiden Hair Fern and crowds her garden.

I saw a Rose garden last June at the country house of a friend, — a house which had been the homestead for a hundred and fifty years, when the question of planting had been solved with little planning or pondering. It was simply a narrow bed around the edge of the semicircular brick wall of a little walled garden or recess; on it were trained Seven Sisters Roses, and all the varied pinkness of Prairie Roses. In the beds blossomed only the few Roses of our grandmothers, and oh! the ineffable fragrance! Damask Roses! Cabbage Roses! Velvet Roses! oh, the perfection of perfume that poured from their pink petals! There were pretty pillars of Madame Plantier Roses at each end of the semicircular bed, each with hundreds of snowy blossoms; and low-growing Moss Roses, and a few ancient Tea-Roses which would fill the spaces with scant bloom later in the summer. Ferns were planted at the base of the brick wall among the Roses, — the common every-day loveliness of New England Ferns. Some of these had fully plumed curves of rich green bursting open; others sent up great quilled ribs of curious form with no leaf-promise in them, and with only a few folded blades thrust out which would in two days open into exquisite fronds. Others were shyly rolled up in cinnamon-colored Catherine wheels and pale brown chrysalis cases like some dull butterfly. Lovely, most lovely, were all these young Ferns among the old Roses. An old stone sun-dial stood on the grass at what was the centre of the circle. Fairy Roses were planted around it, and would bloom when the other Roses were gone. This garden had no magnificent single blooms such as our modern rosarians show to us, and some of the old bushes were sadly gray and broken; but there was the lavish profusion of June to make us wholly happy while it lasted — with no thought for the future months.

Let me close these reminiscences of old-time Roses with these verses by John Russel Hayes, which seem written in the very atmosphere of our grandmothers' garden: —

"O stately Roses, yellow, white, and red,
  As Omar loved you, so we love to-day.
Some Roses with the vanished years have sped,
  And some our mother's mothers laid away
Among their bridal-gowns' soft silken folds,
  Where each pale petal for their sons a precious memory holds.

"And some we find among the yellowed leaves
  Of slender albums, once the parlor's pride,
Where faint-traced Ivy pattern interweaves
  The mottoes over which the maiden sighed.
O faded Roses, did they match your red,
  Those fair young cheeks whose color long ago with yours has fled?

"And still doth balmy June bring many a Rose
  To crown the happy garden's loveliness.
Against the house the old Sweet-brier grows
  And cheers its sadness with soft, warm caress,
As fragrant yet as in the far-off time
  When that old mansion's fairest mistress taught its shoots to climb.

"Enveloped in their tufted velvet coats
  The sweet, poetical Moss Roses dream;
And petal after petal softly floats
  From where the Tea Rose spreads her fawn and cream, —
Like fairy barks on tides of air they flow,
  And rove adown the garden silently as drifting snow.

"Near that old Rose named from its hundred leaves
  The lovely Bridal Roses sweetly blush;
The climbing Rose across the trellis weaves
  A canopy suffused with tender flush;
The Damask roses swing on tiny trees,
  And here the Seven Sisters glow like floral pleiades.

* * * * * * * * * *

"But sweeter far in this old garden close
  To loiter 'mid the lovely, old-time flowers,
To breathe the scent of Lavender and Rose,
  And with old poets pass the peaceful hours.
Old gardens and old poets, — happy he
  Whose quiet summer days are spent much in such sweet company!"

CHAPTER XVII

THE ROSICRUCIANS

"The Rosicrucians are a people I must bring you acquainted with."

—*Dedication to the Rape of the Lock.* ALEXANDER POPE, 1712.

HISTORICAL writing, so say both Carlyle and Ruskin, is made delightful through "disentangling various traceable small threads of relation," by finding interesting "reciprocities and mutabilities," in discovering the beginnings of things. That gratification I have had in tracing the curious and intimate relation of Rosicrucianism to the two subjects of my book, sun-dials and Roses; of tracing, too, these threads through winding by-ways to America. I find in Carlyle's masterly searching out of a Rosicrucian quack, "A Grand Master of the Egyptian Mason Lodge of High Science, Spirit Summoner, Thaumaturgical Metallurgist, Swindler, and Gold-Cook," words literally true of my own searches after similar "Gold-Cooks"; so amusingly like in that I, too, have been unable to obtain a desired book "Which all Librarians make a point of denying that they possess."

The Rosicrucians as a sect have been but little known, though scores of ancient and mystical books and the scattered references in encyclopædias and histories have been followed within a few years with a number not over-satisfactory modern books upon their doctrines. Wild and absurd as were their teachings, they left a distinct trace upon the poetical and legendary literature of Europe. Their beliefs became public in the early part of the seventeenth century, a time when all Europe was seething with religious excitement, when our own country was being settled through these religious controversies. A belief in the baleful possession of foul, malignant spirits, in witches, seemed almost universal, whether these were the spiteful elves of superstitious peasants or the plain devils of the pious Puritans. The new sect claimed that the four elements are inhabited by good spirits which they called sylphs, gnomes, nymphs, and salamanders. Pope says of them in his *Dedication to the Rape of the Lock*: —

> "The sylphs whose habitation is in the air are the best-conditioned creatures imaginable; for they say any mortal may enjoy the most intimate familiarities with these gentle spirits upon a condition very easy to all true adepts — an inviolate preservation of chastity."

The name Rosicrucian was first heard in 1604, though it is said that the sect had existed since the death of the founder, Christian Rosencreutz, in 1484. It is, however, deemed by many careful students very doubtful whether any such person really lived. When it was known that this brotherhood could subsist without eating or drinking; were not subject to disease themselves, and could cure it in others by application of helpful thought; could render themselves invisible; could work miracles; and above all could draw gold and jewels from the earth by incantation, it created great excitement.

It was asserted that Rosencreutz learned the "sublime science" in the East; that while travelling in Arabia he was greeted by some philosophers, called by name at first sight, and claimed by them. From them he learned the secret of prolonging life. It is gravely asserted that he lived to be one hundred and fifty years old, and then died solely because he was tired of living. There were five simple fundamental laws to which the Rosicrucians subscribed: 1. To heal the sick gratuitously. 2. To wear the costume of the country in which they lived. 3. To attend a meeting of the Order at least once annually. 4. To preserve the secret a hundred years. 5. When thinking of dying, to choose a successor. These laws have been observed by Rosicrucian followers to the present day. A few years ago a young friend of mine, who was cheerfully attending balls and dinner-parties in blissful ignorance even of the word Rosicrucian, was accosted in a Parisian drawing-room by an entire stranger who abruptly announced that he had chosen her as his successor in the Society of the Rosy Cross, and would call upon her and explain her duties, as he intended soon to die. She promptly fled from him as from a madman; and after meeting him a second time at a reception, where he gazed upon her without speaking in such an offensively mysterious manner, she left Paris indignantly, simply frightened away; and she has ever since had a timid dread of encountering her Rosicrucian *predecessor.*

Dial-pillar at Wimborne Minster, Dorsetshire.

There came a time in Germany of frantic speculation and investigation on all sides with but little tangible results, for there was little to ascertain; apparently the true Rosicrucians always kept silent. Of course much obloquy came to them; some simply from disappointed curiosity. One writer settled at The Hague after being kicked out of the Society, with the assurance that he would be murdered if he revealed their secrets: "Which secrets," he writes, "I have faithfully kept, for the same reason that women keep secrets — there is nothing to tell." I doubt whether any one living can very lucidly explain and define the Rosicrucian teachings and philosophy. I doubt if the teachers desired lucidity.

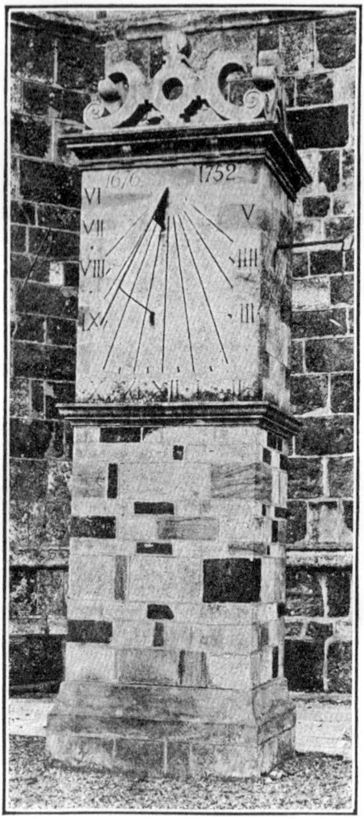

South Dial at Wimborne Minster, Dorsetshire.

They believed in a distinct harmony of the process of nature and the doctrines of religion, and so used chemical terms to express religious truths; they talked of the signatures of things, of the influence of the stars, of magic, of the orders of friendly spirits.

Whether the Rosicrucians were all alchemists, or whether the alchemists were a physical branch of the Rosicrucians, matters little. The art and mystery of alchemy formed an important part of this as of all the mystic religions. When scoffers say in triumph that the Rosicrucians could never have turned base metal into gold, else they would have transformed the world with their wealth, the true "grooms" answer that when they had acquired the power of transmutation into gold, these adepts had ceased to desire wealth.

One alchemist, Sir Edward Kelley, turned a brass warming-pan into silver, and gave away gold wire rings to the value of four thousand pounds at the marriage of a maid-servant, so readily could he make gold. Sir George Ripley gave a hundred thousand pounds to the Knights of Malta for maintaining the war against the Turks—gave it, so it was asserted, because he had found the philosopher's stone in 1470. His works were published by Ashmole. "Robert the Searcher," another great mathematician, could make gold "as easily as he could make salt from sea-water."

Out of all the absurdity and nonsense, the confusion and superstition, of the reports of the Rosicrucians which I have read, there stands out therefrom something which attracts me. I am inclined to suspect that the charm comes largely from the beauty and significance of the Rosicrucian emblem. Old Thomas Fuller in his *Worthies* wrote thus of it:—

> "Sure I am that a Rose is the sweetest of flowers and a cross accounted the sacredest of forms or figures, so that much of eminency must be imparted in their composition."

This simple thought must be shared by many: that the blending of two such significant forms, the Rose and the Cross, must in itself confer dignity on the order. The Rose is almost a universal emblem. A mystic Rose in an allegorical garden is met with in religious traditions of the Orient; there is a Silver Rose in the Garden of Heaven, which is the Brahman paradise. Buddha and the Hindoo god Indra both suffered for robbing a paradisaical garden of a flower. The garden of King Midas, who turned all to gold that he touched, was filled with Roses of sixty petals; the Peruvian Eve of the Garden of Eden sinned not for plucking an apple but a Rose; and the Mexican Eve also gathered a Rose. Into the beautiful symbolic history of the Cross I will not enter. It is a hierogram of even greater antiquity than the Rose. All "persons of sensibility," as the old novel-writers said, have a curious and persistent interest in Rosicrucianism, when once they know of it; or at least a persistent curiosity. Of course I believe that this is also part of the mysterious influence of the Rose, — an influence which exists, though, like all magic, inexplicable, whether this magic be that of the ancient religious mysteries or the simple charm of beauty. We feel this magic of the Rose as we are sensible of the quality entitled "fascination" in our friends; the old Puritan, Cotton Mather, wrote of it, "Of Fascination Man hath more Comprehension than Understanding."

Sun-dial at Talbot, near Bournemouth, England.

A strongly backed derivation of the word Rosicrucian is *Roed* and *Crux*, cross, which I wholly reject through sentiment, and because all the beauty of the story — the hundreds of allusions to the Rose — is thereby lost; the emblematic significance of the Rose as Silence in this most silent of societies, too, cannot be given up.

Rosicrucianism had a unique, an almost comic history in France. The name appeared there in 1623, through a short and mysterious placard which suddenly was in every street in Paris. The name Rosy Cross was as speedily in every mouth. These placards stated that deputies of the Rosy Cross Society Masters were in the city, and were prepared to welcome and teach recruits, but no place was given where they could be seen; no indication of their whereabouts. A burning curiosity was thus awakened in volatile French minds, which was never satisfied. Frantic inquiries in public and private through all channels failed to find any one who had ever seen a Rosicrucian or the teachers. Yet the placards were constantly renewed; and it was told that followers flocked around the teachers somewhere, but became absolutely silent as soon as they became Rosicrucians.

Abuse of the new society was in every speaking mouth; in news-letters, books, pamphlets, and the pulpit was it denounced. The most widespread was a ridiculous bogey book, entitled *Frightful Compact between the Devil and the so-called Invisibles*. Soon the abuse became a crusade of baffled and shocked Roman Catholics against an ultra-Protestantism which was denounced as blasphemy and devil-worship. But it is a poor battle when the fighting is all on one side, and after two years of this noisy but bootless war-

Market Cross with Dials, Woodstock, England. From an Old Print.

fare the Rosicrucian placards vanished as suddenly as they came, and the name even was forgotten for a century.

The Rosicrucian emblems have ever proved so pleasing to the public eye, and the beliefs so alluring to the mystery-loving nature of many people, that they have been often revived by adventurers and swindlers as a romantic surrounding for their quackery and impostures. One of the most interesting of those adventurers was that quack of quacks, Count Cagliostro.

All my notions of Count Cagliostro and the *Affair of the Diamond Necklace* are founded on Carlyle's brilliant essays which, named *Flights*, were printed in *Fraser's Magazine* in 1833. I read them in my early childhood, and in the form in which they were furnished to me they form a curious side glimpse upon the subject of my chapter. These three *Flights* were one of a set of little volumes; the others being *Undine*, a tale entitled *The Holy Hermit* in three volumes; and *Phantasmion*, also in three volumes. This last wonderful book was by Sara Coleridge. It contains some exquisite turns of fancy, but I never knew any one who had read it save myself. These books all had a fine binding, but a greater charm was in the end-papers which, with the marbled leaf-edges, bore a beautiful and singular design; the colors of the prism in little lines formed the background, with a gold Rose pattern stamped upon it,—an heraldic Rose. Not a vestige of these end-papers remain in any of the volumes, the leaf-edges alone tell the story of the pattern. These were the only books I ever wantonly misused; these lovely end-papers of fairy colors were too obviously fit for garments for paper dolls to be left unmolested. The grouping of these books was so significant, and the symbolism of the end-papers so plain, that I am sure they had been thus bound by some one who had, in connection with them, some notion of mysticism, perhaps of Rosicrucianism.

We all know in our own day, in the winters of 1892 and 1893, what nonsense there was in Paris over a revival of interest in this apparently vanished sect. Matinees of the Rosy Cross proved vastly amusing to the volatile Parisians. The high priest of all this was a man prosaically named Josephin, but known to Angels and his followers as Sar Peladan, Grand Master of the Rosy + Cross of the Temple. A description of the meeting of the faithful, "grooms" they are called, does not seem very alluring in the recital, though I think one might get considerable fun at the initiation, when asked by the rigid examiner: "Enumerate thy attractions and thy repulsions." Though Sar Peladan was no end of a *farceur*, he deemed himself the ally of Ruskin, Rossetti, and Burne-Jones, and he was fortunate enough to interest and annex another very great painter, Puvis de Chavannes, whose temperament, as shown in his work, must have made him a ready listener; still, he did not linger long a groom. A less-known but charming French painter, M. Aimé Jean, was also a convert, and contributed to the Rosicrucian Salon his exquisite *Reverie*. M. Khnopff joined the band and gave his painting of beauty and mystery, *I lock my Door upon Myself*. Besides the picture exhibition, lecture courses were given — dull, of course; and plays also. One, *Le Fils des Etolies*, was called a *wagnerie*, and I am sure I know not what that was. Concerts, too, at which a woman, whose

Queen Eleanor's Cross, Northampton. From an Old Print, 1760.

name was seriously given as Mme. Corrylange Mogenboom, performed what were termed "incantations" on the piano.

It is held by the believers in Rosicrucianism that Edward I was initiated into the mysteries of the society in 1296; that the degree of Rose Croix was conferred on him by Raymond Lully (the friend of John Cremer, Abbot of Westminster, and that delightful old English alchemist, Roger Bacon). Edward made the crusade to Palestine and brought back to England with him Guido dalla Colonna and this same Raymond Lully, who coined six millions of nobles for him, as I have recounted at length in another chapter. Lully was a contemporary of Dante and of Arnold of Villanova. All were persecuted exiles. Lully was accused of heresy, and at last took recourse in the language of conventional hypocrisy, and, as did Dante, pretended to be reunited with the Church of Rome.

Many things in Edward's reign can be twisted to hint of Rosicrucianism; for instance, over the door of the Chapter House at York Minster, built in his reign, is this couplet:—

> "*Ut rosa flos florum*
> *Sic est domus ista domorum,*"

thus Englished and rhymed by Thomas Fuller:—

> "Of flowers that grow
> The flower's the Rose,
> All houses so
> This house out-goes."

This house is eight-sided, like the Buddhist cross of the Templars.

Another act of King Edward, which has been claimed as indicative of his knowledge of mysticism and symbolism, was his erection of the wonderful series of funeral crosses to the memory of his wife. Holinshed writes: —

> "In the nineteenth yeare of King Edward, Queen Elianore, King Edward's wife, died, upon Saint Andrew's Even, at Hirdibie near to Lincoln. In everie town and place where the corpse rested by the waie, the King caused a crosse of cunning workmanship to be erected in remembrance of her. Two of the like crosses were set up in London; the one at West cheape, and the other at Charing."

Twelve or thirteen of these splendid crosses were erected. On page 377 is a view of the Northampton cross, copied from an old print of the year 1760. It had at one time four sun-dials on the four faces of the cross.

The cost of these magnificent crosses cannot be known, as so many persons had a hand thereat, it being held both a sacred and a loyal duty. In the upper arches were enclosed four statues of the dead queen. They were the work of William de Ireland, "imaginator" — or sculptor.

This first Edward was certainly a dominant creature with a mind out of the common run of kings of those times. We have heard much of him during these English coronation days, of the Scotch sacred stone with its oracular gifts, vulgarly called Jacob's Pillow, which he bore off incontinently to England, and placed in the Coronation Chair at Westminster Abbey. Its Biblical name rests upon the tradition that this very block of sandstone is one on which Jacob laid his head on that memorable night in Bethel when he saw "angels ascending and descending." From thence it came to Egypt, then to Spain, and then to Ireland, where it was known as the Stone of Destiny. It was borrowed for a Celtic coronation and never returned. Poor Ireland's luck went with it. A succession of Scotch kings were crowned seated upon it, and an old rhyme runs:—

> "Where'er this stone is placed, the Fates decree
> The Scottish race shall there the sovereigns be."

Prosaic geologists report the suspicious fact that it is made of a sandstone abundant near the Scotch town where it was kept so long, and which does not exist in the places named as having prior claims on it. The English regard it with superstitious awe, yet across the broad seat of the chair is carved this inscription in sprawling schoolboy fashion, "Peter Abbott slept in this chair July 4, 1801." It is said that Peter was a Westminster schoolboy, but the date and name sound truly American. One miraculous power of this Jacob's stone was that it groaned aloud if a pretender was seated upon it; but I presume it was speechless with amazement and indignation at naughty Peter's deed.

When the English gave up Rosicrucianism, they took up Free Masonry; the English kings were all interested in it; Charles II is said to have been initiated in France. A dignified follower of both sects was Elias Ashmole, the antiquary, who wrote on the philosopher's stone, and thus became one of Carlyle's "Gold-Cooks." As founder of the Ashmolean Museum, this antiquary must receive some credit and attention. Feasts of astrologers were held, which De Quincey asserts started this English society of Free Masons. Oughtred, the sun-dial maker, whose interesting personality is described in another chapter of this book, was another member. Sir Christopher Wren, who also made sun-dials, was first Grand Master. The Rosicrucians held that all things visible and invisible were produced by the contention of light and shade; and a sun-dial would have been as appropriate an emblem as a Rose and Cross for these English diallers who were so many of them Rosicrucians.

Pillar-dial at Martock, Somerset, England.

The transition from Rosicrucianism to Free Masonry was very easy; I shall not attempt even to indicate it. The Fraternity of the Rosy Cross in England is still chosen from the Masonic body; the same terms are used by the Masons as by the older society. The inquisitive "searcher after mysteries" may gather somewhat of the resemblances, associations, and derivations common to both societies by reading De Quincey's indifferent account of the Rosicrucians; and a much more accurate presentation, Mr. Waite's *Real History of the Rosicrucians*, which I presume is as fair a story of this greatest of all mystic societies as can ever be written.

That curious figure in English history, Friar Roger Bacon, is claimed as a Rosicrucian; as the greatest mind of the thirteenth century, an age rich in great minds, his name certainly would cast honor on any class or society or sect. His manner of thought and his ideas were of the sixteenth century rather than his own, hence he was naturally unappreciated by his contemporaries. His writings were so vast that his biographer said it would be easier to collect the leaves of the Sibyl than the titles even of Bacon's books. He held, as did Frankenstein in our own day, that all the sciences rest on mathematics, even theology; his sketches of geography and astronomy written in this connection are interesting to us because they were the text-books used by Columbus.

Bacon certainly described a method of constructing a telescope, and he is assigned the invention of the barometer; he knew about gunpowder and burning-glasses. He was a firm believer in astrology, in the doctrine of signatures, the philosopher's stone; and he *knew* that *the circle had been squared.* He made some curious prophesies held to apply to steam-engines, balloons, etc. I have been inquisitive enough about Friar Bacon to examine his book, *The Cure of Old Age*, to see what he says of Roses, his loved Emblem, and I find he barely refers to them.

He has as sapless a biography as ever was written. He is not called a Rosicrucian, but, "the vulgar called him a conjurer, and even some Learned men,

Sun-dial at Branksea Castle, Poole Harbour, England.

likewise." In the book are some wholesome rules of health and some shocking ones, but what with conforming everything to the planets, and with wisely quoting of the classics, Friar Bacon contrives to tell as little as I ever knew in a book of a hundred and fifty pages. One Rose reference, of a Rose which would come to life when placed in water, is evidently to the *Rose of Jericho* with which impostor monks often deluded "poor silly women."

There were many remarkable old fellows in England, call them Rosicrucians, cabalists, alchemists, philosophers, what you will, who cannot be set aside as ignorant dupes or wilful liars. Perhaps the most important after Friar Bacon was Dr. Flood or Fludd, who wrote fifteen or twenty great books and was deemed of enough importance to have his works formally refuted by Kepler. He had become infatuated with the teachings of Paracelsus, and endeavored to form on them a philosophy which should prove spiritual and physical birth identical,—a notion which has proved so luring to scores of great thinkers. He was a great mathematician and maker of mathematical instruments, sun-dials, and the like, and he also is said to have invented the barometer. The magic events of his life are interesting to read about, but in the dry-as-dusts which I have ploughed through in emulation of Carlyle, I could not make myself read his long defences of Rosicrucianism. He is described by old Fuller in his *Worthies*, though the doctor knew but little of what he called Rose-Crucians; saying with his usual shrewdness, "Perchance none know it but those that are of it." He wrote somewhat quizzically of Dr. Flood's learning and medical skill: —

"His books written in Latin are great, many, and mystical. The last some impute to his Charity, clouding his

High Matter with Dark Language, lest otherwise the lustre thereof should dazzle the understanding of the reader. The same phrases he used to his patients. And seeing that Conceit is very contributive to the Well working of Physic, their Fancy or Faith natural was much advanced by his elevated expressions."

I think it would interest some of the believers and "healers" of "Christian Science" and the "Mind Cure" to-day to see the frequent hints of similar beliefs among the Rosicrucians. Dr. Flood healed by what he called a "Faith Natural," in which it was asserted that the influence of his mind added to the well-working of his drugs. A total cutting-off of dosing would have been too startling a reform in those days of drugs and dosing. He held that St. Luke was his "physicall and theosophicall patron." Perhaps a diligent reading of the first books of the New Testament with a thought of this discovery in view may show why Luke was chosen.

I have just had a pleasure such as is seldom given to folk of mature years who have ever been greedy readers, undeterred by quality or quantity. Seldom does there remain unread any of the great pieces of literature; but to-night I have read a new and great work, Chaucer's *Romaunt of the Rose*, not literally for the first time, for I recall that I worried through it when a schoolgirl in my study of rhetoric, gulping it down in "required readings," as blissfully ignorant of its meaning as if it were old Sanscrit instead of old English.

Oh, how fine is this *Romaunt of the Rose!* what a picture of youth and chivalry! what a picture of a garden! So plainly worded, so fully described, such "a garden that I love"; as frankly disclosed in every detail as if symbolism had never been invented; and yet it all had a deep meaning; for

Lynn Market Cross, Cheshire, England.

the garden was but a setting to hold the Rose. And the significance of the whole allegory was:—

"*La Rose c'est d'amour le guerdon gracieux.*"

We can glance into the magic crystal so vividly described in the *Romaunt*, and see every detail of the exquisite Rose arbor, the Roser with its varied inflorescence.

The *Romaunt of the Rose* was deemed by the Rosicrucians the allegorical masterpiece of the sect, as it was one of the perfect specimens of Provençal literature. It was written originally about 1282, to satirize the monks; in the double language of Love and Alchemy, it is a complete specimen of Hermetic philosophy.

Throughout Chaucer's works are many veiled allusions which can readily be twisted to alchemical meanings. The poet John Gower, the friend of Chaucer, is another who is claimed as a Rosicrucian. His monument at St. Saviour's, Southwark, shows him crowned with Roses, and with the "three virtues" at his feet.

A Gnostic branch, headed by Lollard with his twelve apostles, united with the followers of Wyclif, and Chaucer was of their number, and left England on account of his belief. Sir Walter Raleigh was also claimed as a Rosicrucian. Spenser's allegorical poetry naturally is claimed by the searchers after proofs of Rosicrucianism; they interpret Una to mean the one true church; the Red Cross Knight can be either the Christian Militant or the Sacred Order of Templars.

The Rosicrucians ingeniously discovered similar emblems and proofs in Dante's *Divina Commedia*. One of the high prophets wrote: "The Paradise consists of a series of Kabbalistic circles divided by a Cross like Ezekiel's pentacle. A Rose blossoms in the centre of this Cross." It was held that in Dante was for the first time the Rosy Cross of the Rosicrucians publicly categorically revealed.

Dante's age was fertile in secret societies and mystical works. If you choose to examine the *Diction-*

ary of Heresies you may learn how many forms a hatred of Rome could take, and how many of these forms were secret societies; the art of speaking secretly, of expressing a thing by means of two meanings, was called grammar; the word is found in variant forms, gramary, glamary, glamer, and the word glamour is therefrom.

Sun-dial on Eastern Gable of Sherborne Abbey.

A wonderful exposition of the secret meanings of the works of Petrarch, Boccaccio, and Dante, of their relations to mysticism, is found in a book written by Gabriel Rossetti, the father of the great artist bearing the same name with the prenomen Dante; this book is entitled rather dryly, *Disquisitions on the Anti Papal Spirit which produced the Reformation: Its Secret Influence in the Literature of Europe in General, and of Europe in Particular.* It is a cultured book, a truly exquisite piece of work, full of instructive suggestion and erudite information upon these poets, and well-termed disquisitions.

Pillar-dial in Market-place, Carlisle, England. From an Old Print.

A distinct impress of Rosicrucian notions is found in *The Rape of the Lock*; indeed, Pope says plainly that he composed his poem "on a new and odd foundation, the Rosicrucian doctrine of spirits." But the sylphs and gnomes and salamanders of the poets' fanciful brain had other pursuits and manners than those of the Rosicrucian philosophy. There have been learned papers written to prove that Shakespeare knew of the existence of the sect and felt its influence. *The Tempest* was written about six years after the outburst of Rosicrucian controversy in Germany; but any thought of a connection between that exquisite creation Ariel and the comparatively clumsy sylphs of the Rosicrucians is unworthy any notice or refutation. Milton's masque of *Comus* is no less zealously claimed for the German sylph-makers, but in vain; the graceful mythology of Greece stimulated the brain of the poet. The masques of the times of James I and Charles I, however, plainly show Rosicrucian influences.

CHAPTER XVIII

THE SUN-DIAL OF AHAZ

"When Joshua Fought Against his Enemies' Force
Bright Sol and Luna sudden stopt Their Course
And Jael's Female Strength had Sis'ra found;
The Stars Assisted in his Fatal Wound,
And Hezekiah's Suit for Life was Done,
Then Ten Degrees Quite Backward Went the Sun."
—*Motto on Dial at The Isle, Shrewsbury, England, 1745.*

THERE lived in America, in New England, in Boston, a much-loved writer, Dr. Oliver Wendell Holmes, who found great and constant amusement in the last lingering years of his busy life, and I believe to some extent throughout the whole of that long and happy life, in the noting of coincidences. These coincidences are simply one of Carlyle's "reciprocities and mutabilities, slight traceable threads," to which I refer in the preceding chapter. It is in that same spirit of pleasure in coincidences that I have noted and shall relate in this chapter the place and relation of the sun-dial and the Rose in a very interesting series of historical incidents beginning with the first dial on historical record, that of Judah in the eighth century B.C., and ending on the peaceful banks of the Wissahickon, in the settlement of Pennsylvania in our own new world.

In 2 Kings xx. 9–11, we read of this first dial: —

"9. And Isaiah said, This sign shalt thou have of the Lord, that the Lord will do the thing that he hath spoken: shall the shadow go forward ten degrees, or go back ten degrees?

"10. And Hezekiah answered: It is a light thing for a shadow to go down ten degrees; nay, but let the shadow return backward ten degrees.

"11. And Isaiah the prophet cried unto the Lord: and he brought the shadow ten degrees backward, by which it had gone down in the dial of Ahaz."

This miracle is told with equal explicitness in Isaiah xxxviii. 8: —

"8. Behold I will bring again the shadow of the degrees which is gone down in the sun-dial of Ahaz, ten degrees backward. So the sun returned ten degrees by which degrees it was gone down."

It is impossible to estimate the attention which has been given to this phenomenon in nature, known as the greatest miracle of Isaiah, not only by Biblical expositors, but by men of science to whom a study of this great reversal of the forces of nature was more attractive and more profound than a search into the more personal miracles of the Bible. I have seen many infinitely learned demonstrations in advanced mathematics to try to prove this miracle possible and also to prove it impossible; and it would be amusing, were it not so solemn, to read the elaborate explanations and reasons given for the possibility of this miracle.

Ahaz, eleventh king of Judah, ruled over his land in the eighth century B.C. Being forced in war by the kings of Syria and Israel, he sought alliance with Tiglath-Pileser II, king of Assyria, who aided him, but in return enforced a heavy tribute and imposed upon him the religious belief of the Babylonish church. The Babylonians were the first people who divided time by any mechanical contrivance; and the rectification of the Babylonian calendar had been made only nineteen years before the accession of Ahaz; doubtless the invention of the pole and of the gnomon were both connected with this rectification. Observation chambers for astronomical research were in use in the East until the middle of the eighteenth century, and "the altars at the top of the upper chamber of Ahaz" which Josiah pulled down were doubtless those observation chambers, which Ahaz had adopted with the Babylonian religion and the Babylonian sun-dial.

It seems a long step from Ahaz and Tiglath-Pileser to America and Benjamin Franklin, but in that long interval appear before us a band of actors to play their parts: the Rosicrucians. One of the most interesting epochs of the existence of that society was in our own country in colonial days. It was a touching one as well, for it marked the decay and extinction of the ancient band. The jaded excitement seekers of Paris of the nineteenth cen-

Horologium Achaz.

tury, the grotesque grooms of Sar Peladan, are not the true Rosicrucians. It is unknown to most of us that there is preserved in Philadelphia a unique relic of those mystics, the most interesting memorial of them that there is now in the entire world. This relic is called the Horologium Achaz, the Sun-dial of Ahaz; in it is performed the miracle of Isaiah, — the shadow is cast backward ten degrees by the refraction of water.

This instrument is shown facing this page. It consists of two finely wrought plates, made of an alloy chiefly of copper and silver; the smaller measures five and three-quarters inches in diameter, and forms the base or standard of the instrument. In the centre is a tiny compass an inch in diameter. There are two encircling bands of wrought and chased work, representing mythological characters and mythical monsters; the under part of this base is even more interesting than the upper portion. It is depicted on page 395. It has a finely engraved and gilded plate, divided into four outer and one middle parts. Two of these are graduated for different elevations, and two contain most interesting panels. One of these represents the king Ahaz sick in bed, while the prophet Isaiah points to a drawing on the wall of the sun and a vertical sun-dial. The other panel apparently figures the twenty-first verse of the chapter of Isaiah which I have just quoted. This verse reads: —

"For Isaiah had said, Let them take a lump of figs and lay it for a plaster upon the boil, and he shall recover."

I cannot refrain from saying that in all my old herbals, arts of simpling, and ancient *Family Chirurgeons* figs are recommended precisely as in Isaiah's recipe, and for the same affliction. In this realistic panel Ahaz is thrusting out his aching leg to be poulticed, while an attendant waits, bearing, apparently, a basket of figs. The fifth and central panel bears a Latin inscription, which may be translated thus:—

> "This semicircular shell explains the miracle of the 38th chapter of Isaiah. For if you fill it to the brim with water, the shadow of the sun is borne backward ten or twenty degrees. Moreover it indicates any common hour of the day, with what is called the hours of the planets."

The upper plate is the sun-dial proper; it is basin-shaped, ten inches in diameter, with flat, movable rim an inch wide. It is shown on page 396. Upon the upper part of the rim are engraved the signs of the zodiac; underneath is this inscription, *Christophorus Schissler Geometricus ac Astronomicus Artifex Augustæ Vindelicorum Faciebat Anno 1578.* The central plate is divided into the different planetary houses; it is about one and three-quarters of an inch in depth. Upon the rim stands a brass figure representing an astrologer, with extended left hand to hold the gnomon, which is, however, now missing.

By filling this shallow basin with water or any transparent liquid, it can readily be seen that the indicated time was advanced or retarded as much as the angle of refraction; thus was the miracle consummated. This instrument was used doubtless in the calculating of nativities, and for various solar observations; and above all for the astrological uses which would make it so valuable to its early owners, the Rosicrucians.

In the name of the maker, Christopher Schissler, we have a clew to its manufacture and history.

Base of Horologium Achaz.

Schissler was a great mathematician in the days when to be that meant much to all men. He called himself a geometric and astronomical master mechanic, and he was an apprenticed brass worker by trade. That he was a skilful artificer, we need only this beautiful and ingenious dial to show. He made likewise a fine quadrant, alleged to be of solid gold, for the Bodleian Library at Oxford, England. This

Basin of Horologium Achaz.

was a Rhenish foot square, and weighed several pounds. It was easy to use gold by the pound when you were wise as to the philosopher's stone.

I am sorry to spoil this tale of the luxury of old-time science by telling that the custodian at the Bodleian Library assures me that the quadrant is only gilt-brass, which any poor simple soul might employ for his instrument. A very good illustration is shown of this interesting quadrant on page 399 of this book, from a full-size photograph taken specially for me at the Bodleian Library.

This instrument is wholly unlike any quadrant I have ever seen; but it is certainly a beautiful example of the engraver's art, which, however, does not show in the reduced illustration. It proves Schissler what he is termed in contemporary biography, — a master brass worker. I have also given in enlarged form, on pages 401 and 403, the panel of engraving seen at the base of the quadrant; for it shows so clearly the tools and belongings of a sixteenth century dialler and alchemist. On one panel is seen, partly concealed by a curtain, a furnace; while at a table sits the old dialler in fur cap and gown with furred sleeves. It is of interest to note his chair, of nearly four centuries ago; that it has claw feet and legs, and back and arms shaped much like the well-made office-chairs of our own day. The table, too, might be a modern dining table. All the instruments are of interest, but I see among them no Sun-dial of Achaz. In the second panel is given the inscription, which is the same as that upon the Horologium Achaz, and a younger dialler in bombasted breeches and jerkin and a smarter hat. The instrument on which his hands rest might be, from its appearance, a modern typewriter; it may be a waywiser. Nor can I tell the significance of the vase of coruscating lines of light which appears in both designs. These two dial-makers, as well as Nicholas Kratzer, and our nineteenth century horologier, John S. Bailey, whose home is so near the settlement of "The Woman in the Wilderness," all have their heads closely covered while at work.

Schissler made much ingenious apparatus and interesting automata for Emperor Rudolphus II; and he made an armillary sphere for his native town, where it was proudly exhibited for many years; but Mr. Sachse, in his fine paper upon "The Horologium Achaz," says it is now missing. He also made the sun-dials on the Perlachthurm at Augsburg, a great tower built as a watch-tower, but now used as a fire lookout. One of these dial-faces can be plainly seen in the picture on page 405. Among other instruments he tried his hand, too, at a waywiser. The Emperor Rudolphus II had a curious waywiser or odometer which is attributed to Schissler.

Let us trace the journey of this unique sun-dial from Schissler's workshop hither. When the Rosicrucians were under social and religious proscription in Germany, they thought and dreamed, as did enthusiasts of every belief at that day, of the new world. In Pennsylvania liberty of conscience was promised to all men, and thither the persecuted Mystics turned their tired feet. Forty was the mystic number of the pilgrims — the "Chamber of Perfection." Six of this band were pastors — one was Zimmerman, and he owned the Horologium Achaz. He was a famous astronomer, a brilliant creature, and he knew, as did all these adepts, the

Schissler's Quadrant; Bodleian Library.

knowledge of living forever; "nathless he died" (as said an old ballad). Sadly was Johannes Kelpius, a young man but twenty-four years old, but of serious nature, made Magister in his stead, and bravely they embarked.

On June 23, 1696, there landed at Philadelphia the ship which bore these forty men, some in pilgrim garb, some in student dress, all in "Out-Landish attire." By tradition, that night, St. John's Eve, members of the band repaired to the highlands northwest of the city, and there the mystic rites of St. John's Eve were first performed in the new world; the blazing boughs were cast down the hill, as had been done since heathen days in the old world, and as was still done till our own day in Pennsylvania. Under the leadership of solemn Kelpius, the pilgrims soon were settled in Germantown, and the society known as "The Woman in the Wilderness" was established.

The brothers soon broke from their vows of celibacy; it was hard for a man to exist as a colonist without a wife. Wives were entreated to come hither, were married in astounding haste, after preposterous courtships; they were bought from shiploads brought hither from England and France simply to become wives for the eager emigrants. Widowers joined in the chase with bachelors with most undignified celerity. A widower of a twelvemonth was far rarer than one who married in six months after the loss of his partner. The Rosicrucian brothers could not withstand the general trend, and soon had some very pretty courtships of as much worldly romance as if they had never seen visions and formed the Chamber of Perfection.

Kelpius was a devout student of the Book of Revelation and the *Morgen-Rothe* of Jacob Behmen. He thought the Quakers too exclusive as to their preachers, and had no sympathy with the severity of Calvin and Luther. For eleven years he led his little band, and it is a wonder he lived so long, for he fasted and prayed in a cave in the earth, suffering "a great cold," which finally ended his days. He believed for many years that he would be translated, as was Elijah, but at the last he told his faithful servant that that happiness was to be denied him. He gave this follower a casket,

Engraved Band on Quadrant of Christopher Schissler.

in which was his magic crystal, and told him to cast it in Schuylkill. It exploded, when it touched the water, with flashes like lightning and rumbles like thunder.

This crystal was the stone of wisdom, a mystic souvenir of the times of Van Helmont, Paracelsus, and Agrippa, with which Kelpius had seen strange sights and had known strange things. It lies in the Pennsylvanian river, waiting, I suppose, a resurrection by some new and, I fear, less picturesque prophet.

Kelpius died in 1704, sitting, so it is told, in his little garden, surrounded by his grieving disciples, and was buried in that garden. It is pleasant to know that memorial tablets to him and to the other Mystics are being placed upon the spots which were their homes and meeting-places.

In the stories of the civilization of the new world there is none so full of charm as that of Pennsylvania. Penn's settlement is replete with pleasant detail and incident, but the story of the later bands of gentle German Christians never fails to touch me. It is uplifting in its faith and trust, in its mutual kindliness, the charity and friendliness of each with all the others. Something of the spirit of these colonists has entered into Whittier's peaceful poem, *The Pennsylvania Pilgrim*. Here are his lines on Kelpius: —

> "Or painful Kelpius from his hermit den
> By Wissahickon — maddest of good men —
> Dreamed o'er the Chiliast dreams of Petersen.
>
> "Deep in the woods, where the small river slid
> Snake-like in shade, the Helmstadt Mystic hid,
> Weird as a wizard over arts forbid.
>
> "Reading the books of Daniel and of John,
> And Behmen's *Morning-Redness*, through the Stone
> Of Wisdom, vouchsafed to his eyes alone.
>
> "Whereby he read what man ne'er read before,
> And saw the visions man shall see no more
> Till the great angel, striding sea and shore
>
> "Shall bid all flesh await, on land or ships,
> The warning Trump of the Apocalypse
> Shattering the heavens before the dread eclipse."

Other religious enthusiasts had come to the Wissahickon before the Rosicrucians, — such believers as the Labadists and Mennonites; and shortly after them followed quickly Moravians, Dunkers, the Ephrata Community — all sturdy Christians and splendid colonists, great housekeepers, prodigious workers.

Here the followers of the Rosy + Cross soon established all the beloved emblems of their secret belief, and carried out all its customs. They had at once, the very first year, an observatory with a chamber like that of Ahaz, — the first regular observatory in the colonies, and possibly the first in North

Inscription on Quadrant of Schissler.

America, — and an ancient telescope, which may still be seen keeping lonely and musing company with the Horologium Achaz over all the changes it has seen. Through this they watched the stars, and thus cast the wondrous horoscopes upon which they made so much of their everyday life depend. For they and their children's children all carried astrological amulets, — small sheets of paper or parchment inscribed with simple horoscopes and astrological signs; pilgrims' tokens, like those borne by the pilgrims and crusaders to Jerusalem. These were worn around the neck of new-born babes, and even placed in the bosom of the dead. These amulets were prepared at midnight in the philosopher's chamber, and assumed on St. John's Eve. They gathered herbs in the dark of the moon, and distilled them with painful care; some of their descendants do so still. They ever searched for the philosopher's stone, and distilled and boiled and stewed and brayed in a mortar for it with as simple faith as they brought to the concoction of thoroughwort and tansy teas. Some of the mystic signets of these simple saints still are in existence,—bits of thin metal, engraved with Rosicrucian symbols. These signets cured disease in man or beast, being pressed on the affected spots, and chanted and incanted over.

Another curious custom which they had was the distribution of printed cards bearing a moral couplet or a verse from the Bible. Two or three hundred of these cards were kept in little boxes and carried by the faithful. Whenever an oath or ill word was uttered in the presence of the casket-bearer he offered one of these slips of paper, chosen at random, to the offender. If he were a brother, he read it carefully and placed it on his tongue. From this arose the grotesque notion that the Pietists *ate* their religion. This custom continued for many years among the Pennsylvania Germans; and these "moral cards" were printed on their presses.

They all looked to the millennium, and some vast sign was to be given on the opening day of the new century after their coming; but they fancied the year

1700 began the century, and disappointment was their lot. Here in our new world the mysteries, rites, and customs of what we now term occult philosophy and esoteric theosophy were practised and carried on by these German pilgrims. It forms to me a strange and interesting picture, the most romantic episode in the planting and growth of our colonies; I love to picture its details. The curious reader may learn in full of the daily manners and religious customs of these various German sects from two massive books of Mr. Julius F. Sachse — namely, *The German Pietists in Pennsylvania*, and *The German Sectaries in Pennsylvania*, for in these two compre-

Perlachthurm, Augsburg.

hensive volumes are gathered most ample details. I like better myself, however, to trace their story in various old books, in copies of old manuscripts, in articles of their household and domestic use preserved in the various collections in Pennsylvania, and above all in visits to the little towns which formed their homes, where still lingers much evidence and exist many of the customs of pilgrim days. In Bethlehem and Zoar and Ephrata the imaginative visitor still may meet the spirit of Conrad Beissel or Kelpius or Zinzendorf in the old streets.

The last magister of this faithful band was Conrad Matthäi; like the prophets of old, he came with sandalled feet and long cloak, and bearing scrip and staff. In his hat he wore a shell like a holy pilgrim, in this case one of the common shells of that vicinity. In Erasmus one says to the Pilgrim:—

> "What kind of apparel hast thou on? Thou art beset with semicircular shells, art covered on every side with images of tin and lead, twined with straw chains and thy arm hath a bracelet of beads."

One faithful follower survived Matthäi—Dr. Christopher Witt, "a very pleasant facetious and plaint old man" who lived to be ninety. A man after my own heart, fond of books, fond of music, fond of flowers, fond of wonders and symbols and emblems and mystic things. When he died he left "an old Virginal," on which he had been wont to play; "a Lot of Old Brass Things," one of these being perhaps none other than the Horologium Achaz; "Belongings Apothecaries ways and Belongings Doctors ways," and an old "Chrinter" whose significance I cannot decipher. He founded a Botanic Garden twenty years before that of his friend Bartram, in whose correspondence with the English botanist Collinson his name often appears. Shrewd old Quakers these were; much of human life is in their letters. They all exchanged slips and seeds and flowers, like any three old maids in a country village, and had some jealousies and slyness over their flowers, too. Collinson wrote of Witt in one flower exchange, "the old gentleman has been too cunning for thee." A pretty picture is shown in one of Bartram's letters, of the twain, Bartram and Dr. Witt, walking happily in the garden; and then talking divinity within the house to some disagreement; and then going out again among the flowers and talking botany to make friends again. Bartram says: —

> "This afforded me a convenient opportunity of asking whether he observed any kind of wild Rose in this Country that was double. He said he could not remember that he ever had. When upon the topic of astrology, magic and mystic divinity, I am apt to be a little troublesome by inquiring into the foundation and reasonableness of these notions, which thee knows, will not bear to be searched into, though I handle these fancies with him with more tenderness than I should with many others so superstitiously inclined, because I respect the man."

Blindness came to him in his last years. Bartram took him through his garden and showed him the gayest new carnations, but he "couldn't tell a leaf from a flower." "A well-spent life will give him consolation, and illuminate his darkness," piously wrote the English botanist.

The scientific belongings of the last of the Rosicrucians came into the keeping of a new and strong light in science, albeit not at all a mystic, Dr. Benjamin Franklin. His love for sun-dials is shown in his introduction of it as a symbol in the first coinage of the new nation. I have also wondered whether the waywiser which the doctor had attached to his chaise, by which, when he was postmaster, he set all the mile-stones on the post-roads — I wonder whether this American waywiser was of the "Old Brass Things" and had been made by Christopher Schissler. It was natural that Dr. Franklin's pet, the American Philosophical Association, should have the antiquated treasures of the Rosicrucians, among them the Horologium Achaz. A good custodian has the society proved itself.

War soon filled the land, and the Revolutionary War ended the existence of this Pietist sect, but the Rosicrucian teachings of Kelpius had been carried from the banks of the Wissahickon to those of the Cocalico at Ephrata. Conrad Beissel had been initiated into a Rosicrucian chapter at Heidelberg, and having thereby lost employment he, with several brother pilgrims, crossed the ocean in 1720 to join Kelpius in his tabernacle. They found the tabernacle deserted, Kelpius dead, and his followers scattered and, alas, nearly all married. Seetig and Matthäi and Witt remained. At the advice of Matthäi, Beissel then helped to found the Ephrata

Community. On Christmas Day, 1723, fires were lighted on the hills and twenty-three members were admitted, six in midwinter being baptized by immersion. Love-feasts and foot-washings soon began; these are still part of the services of some of the Ephrata and other of the German sectaries. The love-feast consists of coffee and unleavened bread, or sometimes lamb soup, to the communicants; while bread and apple-butter is served to visitors. Under

Sun-dial, Church of Our Lady, Munich.

Beissel the sect grew wondrously, and he was one of the earliest patrons of Franklin's printing-press and had strong longings to have Franklin join the community. But I can imagine nothing less to Franklin's taste and nature. As a Christian evangelist Beissel curiously combined mystic theosophy with his teachings in such a way as to be intensely alluring. Some of his disciples carried Rosicrucianism further than he approved; for some returned too close to nature, clinging to the teaching that the Oak furnished the first food for mankind, giving acorns for food and honey-dew for drink, while the rustling of the branches plainly indicated the presence of God. They ate only bread and porridge made of acorn flour, deemed reasonably palatable by hungry brothers. When roasted like coffee, acorns made an excellent drink, and a sour mash of acorns afforded whiskey, since even saints must have that comfort.

Substantial houses were built at Ephrata,—"Brother Houses" and "Sister Houses"; a peculiarity of them was their construction without iron. Wooden pins were used everywhere instead of nails. In cabalistic and Rosicrucian teachings iron was the product of the powers of darkness. In the love-feasts platters of wood were used, and wooden candlesticks were made for the religious meetings; the communion vessels still in use were of wood turned with bronze tools; the linen altar-cloth, even to the present day, after being washed, is smoothed with the wooden flat-irons made and used of old.

CHAPTER XIX

RURAL SAINTS AND PROPHETS

"Die when I may, I always want it said of me by those who know me best, that I always plucked a thistle, and planted a flower where I thought a flower would grow."

— Abraham Lincoln.

IN New England fifty years ago one condition of rural life was universal; and to-day it exists to a considerable extent. This was the personal reserve and repression of feeling of social and domestic life. Not only did this reserve exist between acquaintances and friends (where it had a vast and on the whole a good influence), but it permitted no public and scarcely any private expression of affection or warm interest between lovers, were they married or affianced. Man and wife, whose affection was never doubted in a community where all bore the same reserve, moved throughout life in different though precisely similar circles of revolution — as do double stars. They were never, even when shaken by sorrow or startled by the unexpected, brought into absolute union or moved into any open familiarity or tenderness. Their daily speech had a certain remoteness; they referred to each other vaguely, even when talking among kinsfolk; the wife called her husband "he," or "your father," or "Mr. Stone." Sad it was when a great sorrow like the death of a child found the twain still wearing a mask of reserve which had been fastened on their foreheads by centuries of Puritanism. It seems somewhat anomalous that this should have been the result of Puritanism, since to the early Puritans we certainly owe the first true establishment of the English home and home life. All the sources of history, the letters and journals of that day, prove this true; and those first Puritans displayed a tenderness which had wholly vanished — or rather the display of it had vanished — two centuries later. Read the exquisite love-letters of John Winthrop to his wife — his expressions to others about her. Read of his passion of tears at a public reception when attempting to speak of leaving his home and family to cross the seas. John Eliot, Roger Williams, and the Mathers were as tender as true. In Roger Williams's letters are some exquisite passages of prose like purest poetry telling of his affection for his friends.

Betrothed folk in later New England concealed their betrothal as long as possible; and they were never betrayed in public into any of the affectionate expressions which might be expected — and forgiven them. Nor was applauded the exchange of frequent gifts between lovers save in a few lines of offering, such as delicate food, fruit, nuts, raisins, etc., in the days when these took the place of sweetmeats. Flowers have ever been a true lover's gift; and any lover could show his affection for his sweetheart very properly by assisting her in the care of her flowers, as well as in gifts of flowers. I know one case where the working together in her garden was the only expression of mutual interest indulged in for twenty years by this twain who, nevertheless, loved each other long and deeply.

He was the son of the Presbyterian minister of the town. She was the child of the minister's best friend in his youth. This friend was a dazed, bewildered soul, of some great mental gifts and others of equal weakness, such as the extreme religious atmosphere of New England often evolved. After a childhood and early manhood torn with alternations between deep despair and intense religious exaltation, he finally settled into open infidelity. He had studied, too, to be a minister, and had broken down his health and his hope and his trust by that fierce struggle for an education of one who "works his way" through a New England college. He had had an interval of renewed hope and confidence in God, in which interval he married; but the death of his wife at the birth of Mercy once more unsettled his faith, and left him the prey of every fanatic influence in book or human form which reached the little village which was his home. His crowning offence was that he was moved often to attempt to argue and confute the faith of others. Parson Holmes treated this vagary with silent contempt; but the infidel always could catch a deacon with his bait; and generally the deacon was sadly worsted, for he was not as well fortified for the controversy as the "impious Bible student." The parson would have been a stouter adversary, but he

Sun-dial at Wroxton Abbey.

scorned to fight, and in that was wise. There had been another and a worse thing which darkened this man's life, — he had once attempted suicide. It mattered little in the judgment of the villagers that he had been very ill with a raging fever, and when he ran from his bed and jumped from the open window that it might have been in an access of delirium. He seemed sane enough when they picked him up after his terrible fall, and sorry only at being unsuccessful. This was a finishing climax and a cause for unending odium in a community that still had a law authorizing the burial of a successful suicide at the cross-roads, with a stake driven in his heart.

We can scarcely enter into the profound abhorrence of the minister and his wife, even at the thought that the son whose life they had consecrated to God could in his maturity wish to give that life to the daughter of an unbeliever, and an aggressive, loud-voiced unbeliever, too, such a one as a century earlier would have been stood on Sabbath days upon a block or in a cage, and labelled large that all might read and abhor, "A WANTON GOSPELLER," or perhaps even whipped with many stripes. Their only reason for gratitude was that he had never spoken to the girl of love — but *she knew*. He had walked home with her from singing-school; and they had gone Maying together, driven out to gather the sweet Trailing Arbutus, and that was an open act of keeping company, an hereditary transmission from the old world of May-day customs. His father and mother spoke to him but once, and then besought him not to proceed further in his love-making; and though he made no promise, nor did he cease to walk home with her from singing-school and prayer-meeting, still they knew he spoke no direct word of love to her.

Dragon Gnomon.

Even after the death of both minister and infidel, the widow of Dr. Holmes, simply through her horror of an infidel and a suicide, could keep her son silent as to his love for the daughter of such a one, though he loved the girl far too well ever to heed thought of another sweetheart during all those years. The twain had had one long and untrammelled talk after ten years of silence; it could scarce be called a lovers' scene, though he told her that he loved her and would never marry another. The reserved girl in an agony of plain speech which seemed to her fairly immodest, implored with him for her happiness; she told of her infinite patience,

Hedge of Empress of China Rose.

and she quoted the Scriptures in her favor, and she pleaded that she was and ever had been a consistent Christian, a faithful church-member, and should not be sacrificed for the wrong thinking of another; but still his answer came that he "hated to cross" his mother.

During all those weary years there was one solace for the girl and her lover — her pretty strip of a garden. He worked frequently in it with her, and with no adverse comment of friend or neighbor; even his mother made no opposition. He eagerly secured for her rare seeds and slips wherever he could obtain them, lavishing in this impersonal fashion the affection he should have bestowed directly upon her. And he watched with her their growth and unfolding, lingering over and cherishing a special plant with something of the tender care and thought which should have been bestowed on a child. There was one delicate Rose tree, which he had acquired with much difficulty, and which had needed most constant care. Every known and unknown insect seemed to assail it, and every blemish and blight. John Holmes had spent many an hour bathing with care each delicate leaf and stem with tobacco water, or intently searching for injuring insects. The green fly did not deter him, nor white scale, nor even the "loathly worm," the slug, — these he attacked bare-handed. But this Rose-bush seemed fated to disaster; when a great limb was blown from the elm tree it was borne by the wind and hurled on the Rose-bush. When the strong acid for a solution to assail the Rose-beetle came with misprinted directions,—the word gram instead of grain,—it was this tender Rose-bush which received the withering liquid in its cruel strength. Sometimes they despaired of rearing it, and with sensible reasoning tried to persuade themselves into digging it up and destroying it. But as with a delicate child or pet of any kind, they really loved

Chinese Pedestal for Dial, Floral Park, Long Island.

it the more for the very labor they spent on it—and they loved it, too, though they had never seen it in bloom. Mercy said a little impatiently that she "never expected to see that Rose-bush blossom." One year, by the extraordinary advice of a Rose-growing and avowedly Rose-wise friend, they picked off the buds to try to strengthen the sickly bush; then came a curious blight ever unexplained; then a frost — and in June, too — a frost hoary enough to nip again the promise of bloom. But this year the Rose-bush bore in triumph a beautiful crown of a score of pressing, rounded buds, a glorious promise of fullest beauty; but it was part of the fate of the Rose-bush and of Mercy, that on the day when this rich coronal burst into its fulfilment of glowing ruby bloom — patient Mercy died.

After the death of this gentle girl a singular change came over her lover; he certainly was not insane, but the delicate adjustment of the brain was somehow awry. His friends declared that a slow fever which he had for some weeks had "turned his head." He gave up any careful attention to his farm, raising only enough crops for food; and from the time when the first green leaf opened in the spring until winter snows covered the ground, he seemed to be filled with two ideas, — to help the sick and weakly, and to scatter Roses far and wide. Not the flower! No one ever saw him gather or carry a Rose. Nor did he linger to gaze upon Roses in the gardens of others. His thought was turned toward Rose planting. This certainly should not be deemed an evidence of insanity, when it was the only important evidence there was. It was not the deed, however, but the manner of its doing that made him adjudged a little mad. He would thrust a Rose cutting without words into the hand of some woman as she worked in her little garden, or he would press upon a stranger a dozen Rose hips to be planted. He cared not to know of the growth of these Rose seeds; he simply was filled with a desire to give Rose-bushes to those who wished and, for that matter, to those who did not wish them. But when he gave Roses he also gave many a day's hard work in field or garden to help the widowed or afflicted. Instead of bitterness and resentment over his grief, a great love of humanity filled his soul.

Sun-dial at Saffron Walden, Essex.

In an humble way this man lived up to the fine words of Abraham Lincoln: —

"Die when I may, I want it said of me by those who know me best, that I always plucked a thistle, and planted a flower where I thought a flower would grow."

It was not given to John Holmes to pluck such a great Thistle as slavery, nor to plant the flower of liberty throughout the land; he literally gave simple garden Roses, roots and seeds and cuttings, wherever he thought a Rose might be loved or should grow.

There was no doubt, so said the neighbors who knew him from childhood, that this curious "wheel in his brain" came in some way through uncon-

Sun-dial at Princeton, New Jersey.

scious or insane imitation of a harmless character whom he had often seen in his early life, one "Dr. Jones," who, for many years previous to his death in 1796, wandered through a circle of towns in southern New Hampshire, devoting himself to the spread of choice Apple trees. This man was said to have the first grafted orchard in Hollis, New Hampshire; and while he sold medicinal herbs and herb medicines, he freely gave away Apple scions for grafting. He carried two baskets, one labelled Charity, the other Pity. Dressed in a long plaid banyan, or dressing gown, and a broad-brimmed hat with a flaring mourning weed, Dr. Jones made a queer figure that would be likely to influence the mind of a wondering boy, and later that of a brain-sick man. Holmes had often heard of the love story of Dr. Jones in rhyme, for it was sung by young people in Hollis. The doctor sold his ballad with his "Liberty Tea" and his bunches of herbs and simples. The story was akin to John's own. Dr. Jones had been educated as a preacher and was about to "settle," when he fell in love with a girl to whom his family bitterly objected —

"Because she came of a poor family."

Her father became angry at this scorning of his daughter, and angrily forbade young Jones to visit her.

She promptly pined away and died, while he — so ran his ballad —

"Dressed in black from top to toe
And after that distracted run
And so forever was undone."

He endeared himself to every one in these New Hampshire communities through his gifts of Apple trees and his kindly succor to the sick. His epitaph, in the Hollis churchyard reads: —

"In youth he was a scholar bright
In learning he took a great delight
He was a Major's only son —
It was for Love he was undone."

A similar prophet roamed through what was then the "far West." This singular person was called

Appleseed Johnny, and he was well known throughout Pennsylvania, New York, Ohio, and Indiana for nearly fifty years. During these years he obtained from the Pennsylvania cider-mills enormous quantities of Apple seeds, and made it his life work to start Apple trees in all pioneer communities. At first he bore these Apple seeds in bags on his back, then he had a small horse with leather panniers. Starting out on the Indian trails he planted seeds in the most picturesque spots in the new country; and as pioneers came to settle he had on hand everywhere plantations of young trees to sell to any who cared to buy, or to give away freely. He abhorred grafting, just as he thought it a sin to kill any living creature, even an insect. He mourned deeply killing a rattlesnake that sprang at him; and as he ever went barefoot, it is a wonder he was not often bitten. In his earlier life he wore any cast-off clothing given him, but in later years he had a strange garb,—simply a coffee sack with holes cut for his arms, which were usually bare; and his hat was of pasteboard,—a box. With flowing hair and beard he seemed a prophet. He had a Brahminical horror of eating any but vegetable food, nor would he eat that till he was sure no hungry person could be found who needed it. A Swedenborgian in belief, he expounded his faith and taught goodness and the Spirit of the Lord in every household; and uncouth as was his dress, he was welcome and even beloved. The Indiana farmer owes to him one ill-turn. He became convinced of the usefulness of Dog-fennel as a cure for malaria, and for years he gathered and sowed the seed in vast quantities by every roadside. It is now in these localities a pestiferous weed, under special state legislation for its eradication. He died when seventy-two years old in a farm-house where he stopped to spend the night, and he was sincerely mourned in all farm communities throughout the state. He practically stocked Ohio and Indiana with Apple trees, and was an important factor in the comfortable settlement of both states.

Such characters as these were not unusual in New England; often their "queerness" came from religious excitement. There was the Leatherman, "clad all in Leather," and the "Man with the Scythe." These were perhaps a natural result of the first notions of change in religious feeling in those who had been reared in Puritan rigidity. Nor were these odd creatures as alarming to their neighbors as might be imagined. People were wonted to queerness; they saw and heard such preachers as Lorenzo Dow, the personification of eccentricity. Uncouth in gait and bearing, ill-favored of countenance, deeply pitted with the small-pox, wearing a long ragged beard, "when," says Peter Parley, "nobody among us but old Jagger the beggar had such an appendage," harsh of voice and grotesque in speech, it is a wonder he could have been tolerated; and he was respected and admired. He tells in his autobiography of his strange great-coat which he wished to be buried in, and the two hats which he wore at the same time. His wife Lucy also wore two calashes. He married Peggy Miller after a courtship of barbaric simplicity, he having in his first words to her "made a motion of marriage." After her death he wrote, "My loss was too sensible in contemplating my feelings. Hence my *judgment* dictated a departure from usual custom, and to change my condition again on the journey of life." He acted upon his *judgment* with great celerity. When preaching in Norwich he alluded to his wife's virtues and his loss and closed by calling out, "Is there any one in this congregation willing to take the place of my departed Peggy?" Up rose Lucy Dolbeare, a gigantic woman six feet tall and as broad as she was long, and called out, "I will"—and she did.

CHAPTER XX

A STORY OF FOUR DIALS

"It was intimated to me by divers Worthy Persons as a thing very Expedient that an Explication should be published of the Severall Dyalls here contayned; They judged very Expedient a Declaration thereof should be sett forth."

"A Brief Explication of the Pyramidicall Diall Set up in his Majesties Private Garden at White Hall, July 24, 1669."

— FATHER FRANCIS HALL.

ALL difficulties in the path of the collector, the one who is carried away by the *cacoethes colligendi*, the rage for collecting rather than love of the things collected, add only to his zeal and his delight. Let me then invite his attention to sun-dials, especially if his searches and wanderings be laid in America; for I know no other antique object which is so difficult to find and secure. A friend who is an experienced collector of ancient china, pewter, brass, and such small wares, who knows all the devious ways of collectors and collecting, tells me that in the four years during which her attention has been placed upon the collecting of sun-dials, she has never acquired one. Nay, more than that, she has never seen an old sun-dial in all that time, though her summers are spent in an alert watch in farm-houses or in country lanes and byways; and her winters in frequent prowlings among antique shops.

My Four Dials.

So, therefore, the story of the country acquisition of the four dials shown on this page may prove interesting to collectors, albeit they are no extraordinary tales, but simply accounts of happy wanderings in summer days through the country-side of my beloved New England. The smallest dial came to me through my friend, Addy Dean. I have known this friend about six years; the acquaintance sprang into intimate knowledge and correspondence at first breath, albeit it came about in a most informal manner.

"There's a girl about two mild down the road as has got old traps to sell," answered a New England farmer, whom we hailed in his field. He had come to the roadside willingly, but he imparted his information unwillingly, being bent instead on determined inquiry about us, our identity, our bourne, and our errand We had answered him patiently, recurring ever to our queries about "old blue crockery-ware, old and-irons, pewter porringers, and maple bureaus." "What is her name?" we asked briskly, meaning to drive on and trust to fortune and some one else to find "the girl." "Ye drive down the road to the school-house; then turn to the left and go to a big Oak tree; go down that lane and turn in on the River Road to a driftway —"

Shield-shaped Vertical Dial. Made by John S. Bailey.

"Isn't there a straighter road?" we interrupt. There was a straighter and simpler way, of course; country folk delight in complicated instructions for reaching a destination. "You'll know her when you see her; she's about my age. We went to school together." Of his age! He might have been sixty or eighty or a hundred, as he leaned on his hoe; for his face was so streaked with layers of dirt and honest sweat of the brow that he seemed not to have been born and grown in human manner, but to be formed in geological strata. The Dean Farm proved to be remote indeed from the road,—so isolated that from it no other house could be seen, though there were neighboring farm-houses in sheltered hollows and behind near woods. A silence like that of the enchanted palace in the story of the Sleeping Beauty lay upon the house as we drove into the dooryard in the hot sunlight; every door and window was closed and even locked; no farm "beasts" were around, no friendly cackle of hens and rooster was heard, no bark of dog, no sign even in the air of the ever present hog. A vigorous pounding on the door-knocker wakened, as in the fairy tale of old, the sleeping kingdom. A great dog bounded around the house corner; a brood of chickens followed their disturbed mother from the barn-yard; from a tiny window in an ell came the voice of the princess, "Here's the key," and it fell from the window at our feet. "Go in and see things," she added, in cheerful welcome. It is one of the surprises of New England farm-life that the woman who is housekeeper and houseworker, be she wife or, as in this case, sister, can have time in the afternoon for a comfortable nap. She needs it, of course, having risen early and worked hard; the wonder is that she gets it. From the barn-doors appeared the princes,—the Dean boys,—who had evidently been napping too, on the hay. They gave one glance at us, returned to the barn, and reappeared with a wooden peck measure half full of pears, offered to us with the princely hospitality of a New England farm that gives to you the best of whatever it has. In this splendid "dish of pears" I discovered to my delight that most luscious of all pears, the Flemish Beauty, which I had not seen for years. Two great trees of the variety grew in our garden. They were of most vigorous growth, and the fruit of extraordinary perfection. The Flemish Beauty is liable to a curious warping and splitting, turning black in these splits and in mottled spots, and thus not becoming over inviting in appearance, though its flavor is scarcely injured thereby. Our Flemish Beauties never were split or spotted or blackened; and in looking back upon their growth I can recognize why they were so perfect. They grew in the evil germ-bearing days of open sink-drains; the water from our kitchen must have proved a con-

Sun-dial on Ely Cathedral: "The Lantern."

stant supply of moisture and richness for these Pear trees, which stood on either side of the opening of the drain.

Since that September day when the Dean brothers gave us of their bounty, I have had Flemish Beauty pears from their trees each autumn. When Addy Dean appeared, and in a surprisingly short time, I saw why she was termed a girl. By the testimony of the stratified farmer she must have been sixty years old; but her face, with the fine pink bloom of its cheek, was that of perennial youth. She was thin to a degree, and clad in ancient garments which bore no pretense of modern reshaping. A tucked barège of green and white was worn with a purple sprigged lawn waist of a style of forty years since; its shoulder-seams drooped halfway to the elbow in comic resemblance to a recently revived mode. A tiny fringed black silk cape hung around her thin shoulders. Her pale hair, pale with that curious faded look assumed by golden locks when mingled with silver, was so trimly brushed and dressed that we wondered whether she could have been napping on a Japanese neck-pillow.

Addy Dean is a type of New England life which, I fear, will never be found in generations to come; a life of absolute dignity, even in comparative poverty. She lives with her brothers in their hundred-year-old house, with scant comfort, poor fare, and self-respecting independence, on a farm whose soil throws rocks up to view far more readily than corn. In the kitchen of that farm-house has never stepped, I believe, the foot of hireling servants — not even country help. Perhaps in strenuous time of birth or death some nurse of the country-side may have spent a few days, and been given in payment a bag of meal, a ham, a barrel of potatoes, or a jug of cider-brandy; but it is far more likely that all such offices have been performed by kindly neighbors or kinsfolk. Every detail of housework has ever been done by the women of the household, and in past days they made much butter, too, for sale. Addy Dean is shrewder than her grandmother. To raise eggs to sell is far less work to the farm-wife than to make butter; and selling antiques is more profitable than either. She was clever enough ten years ago to discover, even on that isolated farm, through the queries of the collectors for antique dealers, that there was an opening for her to make money, and she has, as the old books say, "improved" it. Addy Dean was so antique in dress and so simple and direct in her speech that her letters were a great surprise to me as well as a great delight. She has ample fund of quotation and comparison, and shows thereby goodly reading, though she lives six miles from the village library, and two of these six miles are private ways, one mile a driftway through the fields, and never broken out in winter by the town. Winter is, of course, her only reading time. I noted in her house a surprising number of magazines and weekly papers, possibly not of the latest dates, which matters little — far more magazines, I am sure, than are read by the city dweller in general; far more than I read myself. Her spelling is perfection, like nearly all New Englanders of her generation, whatever that may be — I am setting her age as sixty. She was educated at the district school, which was one of the unique microcosms of our new-world life, the town-meeting being another. I doubt whether the scholar received much more personal attention than in our graded schools to-day, but the district school somehow afforded a happier education than is now given; for a good

Ancient Sun-dial(?). Owned by Author.

"all-around start in life," with an education enabling one to write a good letter and to enjoy a good book, I know nothing that can at all compare with the "schooling" of the district school. It had its faults, but they were not so great for their place and time as are the faults of the present graded school. The district school turned out such pupils as Addy Dean; doubtless nearly all children are better scholars, further "advanced," in our public schools to-day than she was at their age. But what I question is, will they be as good scholars as she is when they are her age? I am sure they cannot love their schools as she loves the memory of hers. They will have little to remind them of it. She has a sorely worn dozen of school books — battered as would be school books studied by five or six scholars in succession. But there they are, to jog her memory if she opens them, not only to recall a bit of geography or grammar, but to recall district school and scholars through the scribbled lore on fly leaves and page borders. Nothing can be to me more forlorn, more like a convict system, than the latter-day fact that scholars do not own their school books — these are the property of the city. The child never "minds his book" save in school hours, and never sees it when he leaves.

No child of to-day will ever, forty years from now, show with almost tender reminiscence, a "thumb-paper" which has chanced to remain in her "Reader" since her school days; a thumb-paper being — oh! you poor, unfortunate child who owns not your own school book — a square of stiff letter-paper or possibly colored paper, folded in a certain fashion to protect the lower portion of the leaf from "pricking thumbs." This "thumb-paper" of Addy Dean was of ruled blue foolscap and had as an appurtenance a long bit of linen thread, the end of which was brought around the outer book-back and tucked in securely between the leaves; this was to hold the "thumb-paper" in place, and in this case it had held it for half a century. In these school books of Addy Dean I found with glad recognition all the half-forgotten fly-leaf lore of my own childhood in Worcester, — the warning rhymes and set border drawings of school children; the Preface legends, and the coin-tracings, or rubbings, which we sometimes cut out and used as money in our pin-stores, poppy-shows, and grocery-stores.

Addy Dean has in the village a place where she can display any unusual piece of furniture or china which she has found; it will in the summer-time catch there the eye of the summer visitor, and thus tole him or her to the isolated Dean Farm and further purchases. We are apt to fancy that we secure great treasures if we purchase them from lonely farm-houses. This show room is in the house of a widower, a friend of the farmer who first told us of her. We suspect that the widower is an old admirer; he, too, always speaks of her as a girl. She had no sun-dials when I first visited her home. It was nearly a year before she could send me this one. And I tell, as an example of the dignified way in which this woman performs her duty in her day and her sphere in life, that ere she sent it to me she asked permission to have a copy cast from it, "to give to the children in the public school." On further inquiry I found that she had made for the school of the little town a very good collection of unusual objects which would interest and instruct the scholars, — Indian arrow-heads, curious utensils, old farm tools, carpentering tools, and a few old manuscripts. As a proof of the absolute disappearance of the sun-dial in many localities, let me state that this sun-dial is the only one Addy Dean has ever found for me; and when it was displayed to the scholars in that school, there was not a child present who knew what it was, or who had ever seen a sun-dial; and I am not sure that there was one present who had ever heard of one.

Sun-dial at Didsbury, England.

There may be one special reason for this: Addy Dean lives in the valley of the upper Connecticut, and when the towns were settled there, about the time of the Revolutionary War, clocks had already become cheap and common. Thrifty settlers, and it was such who went there, and who left such descendants as Addy Dean and her brothers,—these careful men managed to carry a small clock, even when the journey was made on horseback.

In our New England home-gardens and home-meadows, and I am told also in yards and gardens in the Middle and Southern states, there grow trees unclassified of botanist, unnamed of scientist. It matters not their fruit, their flower, or their cone or nut, it matters not the shape of their leaves nor the spread of their branches; but be they broad of trunk or bole they are right, and by their *roots* ye shall know them. Any tree that runs into the earth wide-spreading or curiously gnarled roots that extend out a little protected enclosure or circuit, this tree is the Children's Tree, and those roots form for the children their "cubby-house." This word *cubby-house*, universally used by children, is one of those obscure and universal terms so tempting

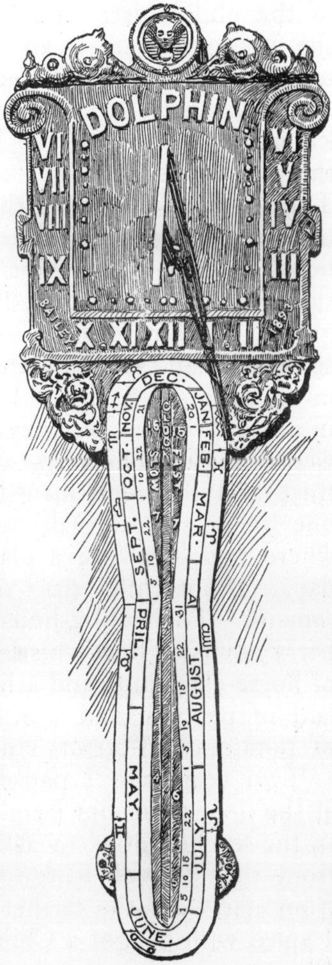

Sun-dial, showing Mean Time.
Made by John S. Bailey.

to the philologist. The word *cub*, in the sense of the young of various animals, has always been contemptuously applied to servants, and, in the eighteenth century, to the assistants in hospitals, now known as "interns." "Each surgeon shall have three cubs as helpers," says an old hospital rule. The word *cubicle* (from the Latin *cubare*) seems to be little used save for the little cubby-holes of our boys' schools, — such, for instance, as the fifth form sleeping-apartments at St. Paul's School. I am always glad to find Dr. Holmes using the word cubby-hole, as he does all New England words and phrases, even in his serious writings. I have no doubt he played in a cubby-house, for they were beloved of boys as well as girls. The boys' cubby-houses showed distinct and different furnishings. I well remember one under a great Oak tree in sunny Narragansett, where two loved boys played many a long summer day. Of one child, alas! only such happy memories remain. This cubby-house held pop-guns of elderberry stems, willow whistles, corn-silk cigars, strings of horse-chestnuts, and a little farm with stone walls laid of pebbles, and wonderful farm animals made of potatoes and carrots stuck with wooden legs.

Two years ago I paused one summer afternoon at the door of an old farm-house and walked around to the kitchen door to ask for "a drink of water" from the old well whose well-sweep had lured me from afar. At the farther end of the kitchen yard I spied with delight a Children's Tree, — an ancient Pine tree; and in the cleanly circuit of its roots, on its needled floor, was set a sight to thrill the happy memories of vanished years. There was all the old familiar stock and store of trash which is transmuted by childhood's subtle alchemy into unutterable treasure: milkweed pods, acorn cups, cleft peach-stones, rose-hips, and that ever present furnishing of cubby-houses, — broken china. American children gather their treasures precisely as is told of the children of the ancient Britons, "pots-herds or broken glasses or bones half burnt, or lime, or plaster"; ground-up and calcined bones make wonderful flour for childish storehouses. We had pretty dolls' tea-sets of turned wood and of real Chinese porcelain, Lowestoft, for which we cared little; there was no imagination about them, and we could not take them out-of-doors and handle them freely.

I noted a few bits of old blue and lustre ware in this wayside cubby-house which promised ancient wares within-doors in the mother's pantry and cupboard, which we might be able to purchase if she "cared to dispose of them" (never to sell them), when my attention was fixed by the little circular table of pewter about seven inches in diameter on which the choicest bits of the child's china were carefully displayed. Like a "table-bord" of ancient days, it seemed to be propped up from beneath by crossed wooden sticks like saw-horses which served as table-supports or trestles.

We had passed the children coming home from school, and even now they were racing down the home lane. So we waited, gazing rather sadly at the cubby-house treasures, until the older girl had run in to us, and at our request carefully removed the china and pulled up and turned over the pewter dining table, displaying to us on what had been the under side the gnomon and lettering of an old sun-dial.

Sun-dial in Garden of William Robinson, Author of the *English Flower Garden*, Gravetye Manor, Sussex, England.

She had found it in the attic and father thought it was part of a water-clock; he had read of such things, but he could not think how it could run. She was overcome with delight at the proposal to exchange her pewter table, her clepsydra, for a pretty travelling-mirror which our dressing-case contained; and the second sun-dial in the group, the one with a circular base, is now in my collection, to remind me ever of that happy summer day and the little cubby-house under the Pine tree.

New England houses are, in general, frankly open in aspect, having no hidden meaning, no suggestion of sentiment. On the Ridge Road stands a house of unusual bearing, unlike its fellows in the old "South County,"—a house which suggests even to the careless traveller something unusual, remote from everyday life and experience. It is so densely fronted with black-shadowed Cedars that you scarce see it in passing; and a single grand old Fir tree, a remnant of the forest primeval, overhangs the Cedars and the lowly roof-tree; shadows of Cedars and Fir tree mantle back into a stately, reserved Pine forest close in the rear. The front windows of the lower story are heavily shuttered behind the Cedars, but the scant half story above has blinking windows at the house ends under the overhanging roof. The house looks like a creature reflecting and enduring in silence. It stands near the cross-roads and is not far from the old post-road, which went up hill and down dale, but is never used for business drives to-day; it shows by these two facts in its placing that its builder had social instincts and interests, that he wished to know his neighbors and the passing world. There is a family burying-place at the end of the garden, enclosed by more rank, darkling Cedars and a sister Fir tree. A row of slate headstones show by their scant inscriptions that a family of brothers and sisters here lie close together. On the grave of most recent date (though that date is 1830) and over the neighboring headstones trails a choked Damask Rose. It is sadly unkempt, full of its own dead branches, and a tangle of Fir cones and many twigs and branches of broken Cedar.

Nearly a year after the death of the last sister there suddenly appeared in the village, unacquainted, and to the surprise of all the curious neighbors, a distant cousin of the family, an elderly woman named Johns, who quietly settled before "the Squire" her claims to ownership of the house, and established herself therein.

Since he was as reticent of nature, nearly, as she was, the curiosity of the neighbors was never satisfied. The lower shutters were then closed, and locked within, and never opened during her life-time,— an action typical of the reserve of her nature; and she thereafter lived in the rooms in the upper story. She was a weaver; I have told her life (in the scant details known to old residents in the South County) in the chapter entitled Narragansett Weavers, in my little book of stories of *Old Narragansett*. There were many hand-weavers in the South County, where wool-weaving by machinery has since been so sturdily established; more than elsewhere in New England. There is one weaver still, William Rose, who weaves yearly on his hand-loom scores of bed-coverlets and hap-harlots, and other weavers who weave rag carpets only. There were in older days many weavers who worked in their own homes, spinning their own wool into yarn and weaving it into cloth to sell, or weaving the thread and yarn brought them by their neighbors; and there were travelling weavers also, who "whipped the cat" from house to house, working for a few shillings a day and their "keep." Sometimes they brought their looms and set them up and spun for a month or two — as when making a wedding outfit.

There were women weavers, too, in plenty, and Mistress Johns joined their number. She soon excelled them all in quality and quantity of her work, and therefore had work in plenty. I suppose it was her silence as well as her singular habit of stopping when at her work, and sitting for hours motionless, which soon gave her the name of being a witch. It often angered the housewives when they were in a hurry for a piece of homespun for the boys' winter garments, to see her sit silent for hours, peering strangely into the loom, but they feared her witchcraft and said nothing; and when night came, all would leave her in the loom-loft with candles or primitive lamps, silent still. But at midnight the family would hear a low, half-toned clapping of the loom; not a loud bang, bang, as of honest weavers, but a dim sound of some one — the Old Boy — of course, helping the witch out on her stint. So at the week's end there was always more linen ready for bleaching, more homespun cloth on the roll, more yards of rag

Sun-dial and Porch of Church, Grateley, Hampshire, England.

carpet ready for sewing than could be turned out by any man weaver in Narragansett.

Therefore, though it might be hitching up with the Devil, even the parson's wife employed her, and every household gave her work in plenty. She never ate with her employer, even when she stayed half the night, nor did they see that she brought food with her; and she never talked, save to learn of her work; and no neighbor ever entered her door, for she was seldom there save after nightfall, anyway. She lived to great old age, the last itinerant weaver in Old Narragansett. There came a time when, after a heavy storm, she did not appear as bidden in a gay household where preparations for a wedding were under way, and where she was to weave rag carpet for the bedrooms of the bride's new home. A kindly little tailoress, learning this, went across fields with a hand-lantern after nightfall, and entered the dark house, and climbed the narrow staircase; a poky thing to do, and a difficult one, for the tailoress was old; a brave and Christian-like deed, too, for she was a timid and superstitious soul. She found the weaver-witch dead in her bed, with the rain blowing in through the broken roof, which had partly blown off in the storm. It was a sad sight, and one that throughout her life ever terrified the kindly little creature.

The following August a visit was made to the "Witch's House" to see if aught of interest or value remained in the house. The end windows had been broken by stones thrown by marauding boys, and spring rains and summer suns had freely

Sun-dial at Chastleton Manor, Oxfordshire, England,

entered through window and roof. And the witch's bed on which she died, a sack filled with straw of mouse-barley with some spikes of grains attached, had sprouted and grown through the coarse hempen bedtick, and thus her bed was as green as the grass over her unmarked grave under the garden Cedars.

I have half a score of her loom-shuttles, and some of her loom-spools, a raddle (or rake), a sley, some niddy-noddies, — all these are portrayed in my book *Home Life in Colonial Days*. And I have her sun-dial, it is the third in the illustration on page 427. A primitive little dial, it served her well through many years of honest work and isolated life; for she had no other timekeeper.

The fourth — the largest sun-dial — is not in very good favor with me at present. I bought it from an ingenuous farmer on The Boston Road. Black and dirty, it was cemented to a tree stump in his kitchen garden, and seemed as old as the house, which he said was a hundred, and which I *thought* he said was built by his grandfather. I was much delighted when I purchased it; but just as I hurried away from the farm kitchen he displayed to us several pitchers of silver lustre and half a dozen blue willow-ware plates which he would be willing to part with. We could only glance at them, as we had scant time to drive two miles to catch the local train, but by that fleeting glance those pieces of old crockery certainly looked brand new. At the station I had one minute to interview a stage-driver. "How long has Ellis lived on his place up the road?" I said. "About a year," was the answer. "Hasn't he a sun-dial for sale?" I venture diplomatically. "Don't know as he has; he sold it last week."

Now if he sold his sun-dial last week, from whence came my dial? My friends believe the dial is an old one, but I think Farmer Ellis is a broken-down city dealer with an attic full of new dials cast in some old mould or stamped with some old die, ever ready to replace the recently sold one in the kitchen yard.

INDEX

Other TUT BOOKS available:

BACHELOR'S HAWAII *by Boye de Mente*

BACHELOR'S JAPAN *by Boye de Mente*

BACHELOR'S MEXICO *by Boye de Mente*

CALABASHES AND KINGS: An Introduction to Hawaii *by Stanley D. Porteus*

CHINA COLLECTING IN AMERICA *by Alice Morse Earle*

CHOI OI!: The Lighter Side of Vietnam *by Tony Zidek*

CONFUCIUS SAY *by Leo Shaw*

THE COUNTERFEITER and Other Stories *by Yasushi Inoue; translated by Leon Picon*

CURIOUS PUNISHMENTS OF BYGONE DAYS *by Alice Morse Earle*

CUSTOMS AND FASHIONS IN OLD NEW ENGLAND *by Alice Morse Earle*

DINING IN SPAIN *by Gerrie Beene and Lourdes Miranda King*

EXOTICS AND RETROSPECTIVES *by Lafcadio Hearn*

FIRST YOU TAKE A LEEK: A Guide to Elegant Eating Spiced with Culinary Capers *by Maxine J. Saltonstall*

FIVE WOMEN WHO LOVED LOVE *by Saikaku Ihara; translated by William Theodore de Bary*

A FLOWER DOES NOT TALK: Zen Essays *by Abbot Zenkei Shibayama of the Nanzenji*

FOLK LEGENDS OF JAPAN *by Richard M. Dorson*

GOING NATIVE IN HAWAII: A Poor Man's Guide to Paradise *by Timothy Head*

HAIKU IN ENGLISH *by Harold G. Henderson*

HARP OF BURMA *by Michio Takeyama; translated by Howard Hibbett*

THE HAWAIIAN GUIDE BOOK for Travelers *by Henry M. Whitney*

HAWAIIAN PHRASE BOOK

HAWAII: End of the Rainbow *by Kazuo Miyamoto*

HOMEMADE ICE CREAM AND SHERBERT *by Sheila MacNiven Cameron*

HOW TO READ CHARACTER: A New Illustrated Handbook of Phrenology and Physiognomy, for Students and Examiners *by Samuel R. Wells*

INDIAN RIBALDRY *by Randor Guy*

IN GHOSTLY JAPAN *by Lafcadio Hearn*

JAPAN: An Attempt at Interpretation *by Lafcadio Hearn*

THE JAPANESE ABACUS *by Takashi Kojima*

THE JAPANESE ARE LIKE THAT *by Ichiro Kawasaki*

JAPANESE ETIQUETTE: An Introduction *by the World Fellowship Committee of the Tokyo Y.W.C.A.*

THE JAPANESE FAIRY BOOK *compiled by Yei Theodora Ozaki*

JAPANESE FOLK-PLAYS: The Ink-Smeared Lady and Other Kyogen *translated by Shio Sakanishi*

JAPANESE FOOD AND COOKING *by Stuart Griffin*

A JAPANESE MISCELLANY *by Lafcadio Hearn*

JAPANESE RECIPES *by Tatsuji Tada*

JAPANESE TALES OF MYSTERY & IMAGINATION *by Edogawa Rampo; translated by James B. Harris*

JAPANESE THINGS: Being Notes on Various Subjects Connected with Japan *by Basil Hall Chamberlain*

THE JOKE'S ON JUDO *by Donn Draeger and Ken Tremayne*

THE KABUKI HANDBOOK *by Aubrey S. Halford and Giovanna M. Halford*

KOREAN FOLK TALES *by Im Bang and Yi Ryuk; translated by James S. Gale*

KWAIDAN: Stories and Studies of Strange Things *by Lafcadio Hearn*

LET'S STUDY JAPANESE *by Jun Maeda*

THE LIFE OF BUDDHA *by A. Ferdinand Herold*

MODERN JAPANESE PRINTS: A Contemporary Selection *edited by Yuji Abe*

MORE ZILCH: The Marine Corps' Most Guarded Secret *by Roy Delgado*

PHYSICAL FITNESS: A Practical Program *by Clark Hatch*

POO POO MAKE PRANT GLOW *by Harvey Ward*